Depleted Uranium Aerosol Doses and Risks

Summary of U.S. Assessments

M. A. Parkhurst[a]
E. G. Daxon[b]
G. M. Lodde[c]
F. Szrom[d]
R. A. Guilmette[e]
L. E. Roszell[d]
G. A. Falo[d]
C. B. McKee, *Project Administrator*[c]

Prepared for the US Army by Battelle under the
Chemical and Biological Defense Information Analysis Center
Task 241, DO 0189, Aberdeen, Maryland

---

[a] Battelle, Richland
[b] Battelle, San Antonio
[c] Battelle Eastern Science and Technology Center
[d] U.S. Army Center for Health Promotion and Preventive Medicine
[e] Los Alamos National Laboratory

Battelle Press
Columbus • Richland

DISCLAIMER

This report is a work prepared for the United States Government by Battelle. In no event shall either the United States Government or Battelle have any responsibility or liability for any consequences of any use, misuse, inability to use, or reliance upon the information contained herein, nor does either warrant or to otherwise represent in any way the accuracy, adequacy, or application of the contents hereof.

For online electronic copies, please refer to www.deploymentlink.osd.mil/, or for hard copies and CDs, please refer to www.battelle.org/bclscrpt/Bookstore/default.cfm.

The Chemical Warfare/Chemical and Biological Defense Information Analysis Center is a Department of Defense Information Analysis Center operated by Battelle Memorial Institute and administered by Defense Information Systems Agency, Defense Technical Information Center (DTIC-AI), Fort Belvoir, VA 22060-6218.

Prepared for the U.S. Army for Health Promotion and Prevention Medicine and the U.S. Army Heavy Metals Office

LIBRARY OF CONGRESS CATALOGING-IN-PUBLICATION DATA

Depleted uranium aerosol dose and risk : summary of U.S. assessments / M.A. Parkhurst . . . [et al.].
   p. cm.
   "Prepared for the U.S. Army by Battelle under the Chemical and Biological Defense Information Analysis Center Task 241, DO 0189, Aberdeen, Maryland."
   Includes bibliographical references and index.
   ISBN 1-57477-148-5 (alk. paper)
   1. Uranium—Health aspects. 2. Uranium—Environmental aspects. 3. United States. Army—Weapons systems. 4. United States. Army—Sanitary affairs. 5. Projectiles. 6. Tank warfare—Health aspects. I. Parkhurst, M. A. II. Capstone DU Aerosol Characterization and Risk Assessment Program. III. Chemical Warfare/Chemical and Biological Defense Information Analysis Center (U.S.)

RA1231.U7D47 2005
363.17'99—dc22

2005042817

Copyright © 2005 Battelle Memorial Institute. All rights reserved.

This document, or parts thereof, may not be reproduced in any form without the written permission of Battelle Memorial Institute.

Printed in the United States of America

Battelle Press
505 King Avenue, Columbus, Ohio 43201-2693, USA
614-424-6393 or 1-800-451-3543
Fax: 614-424-3819
E-mail: press@battelle.org
Website: www. battelle.org/bookstore

Participating Organizations

Acknowledgments

The *Capstone Depleted Uranium Aerosols: Generation and Characterization Study* (Attachment 1 and 2) was conducted under the auspices of the U.S. Army Heavy Metals Office. It was financially supported by the Office of the Special Assistant for Gulf War Illnesses, Medical Readiness and Military Deployment and by the U.S. Army. The Depleted Uranium Research-Integrated Process Team (DUR-IPT) was charged with project oversight. Mr. Michael Los, a DUR-IPT member from the Assistant Secretary of the Army (ASA) for Acquisition, Logistics and Technology, US Army Heavy Metals Office, was the Army lead for the project and the point of contact between the DUR-IPT and the project team. Additional organizations directly contributing to its performance included the Army Surgeon's Office, the PEO-GCSS, USACHPPM, MEDCOM, AMC, ARDEC, ATEC, OSD, and others.

The team responsible for planning, implementing, analyzing, and reporting the results of this study are the following:

- M. A. Parkhurst, Principal Investigator, Pacific Northwest National Laboratory
- F. Szrom, U.S. Army Center for Health Promotion and Preventive Medicine
- R. A. Guilmette, Lovelace Respiratory Research Institute, currently of Los Alamos National Laboratory
- T. D. Holmes, Lovelace Respiratory Research Institute
- Y. S. Cheng, Lovelace Respiratory Research Institute
- J. L. Kenoyer, Pacific Northwest National Laboratory, currently of Dade Moeller & Associates, Inc.
- J. W. Collins, U.S. Army Center for Health Promotion and Preventive Medicine
- T. E. Sanderson, U.S. Army, Aberdeen Test Center, Aberdeen Proving Ground
- R. W. Fliszar, U.S. Army, TACOM-Armament Research, Development, and Engineering Center
- K. Gold, U.S. Army, TACOM-Armament Research, Development, and Engineering Center
- J. C. Beckman, U.S. Army, Aberdeen Test Center, Aberdeen Proving Ground
- J. A. Long, U.S. Army, Aberdeen Test Center, Aberdeen Proving Ground

The independent peer reviewer science group, who contributed richly to project guidance and draft document review, was led by Dr. Roy Reuter and included Dr. Paul Baron, Dr. John Doull, Dr. Rogene Henderson, Dr. David Hoel, Dr. Morton Lippmann, Dr. Paul Strickland, Dr. Arthur Upton, and Dr. Wes Van Pelt.

The *Human Health Risk Assessment of Capstone Depleted Uranium Aerosols Study* (Attachment 3) was conducted under the auspices of the US Army Center for Health Promotion and Preventive Medicine, LTC Mark A. Melanson, PhD, CHP, Health Physics Program Manager. Mr. David P. Alberth, Task Sponsor Representative (Health Risk Assessment for Exposure to Depleted Uranium—CBIAC Task 241) and USACHPPM Master Consultant, provided programmatic guidance and management of the technical review process.

The team responsible for planning, implementing, analyzing, and reporting the results of this study are the following:

- R. A. Guilmette, Los Alamos National Laboratory
- M. A. Parkhurst, Battelle, Richland
- G. Miller, Los Alamos National Laboratory
- F. F. Hahn, Lovelace Respiratory Research Institute
- L. E. Roszell, U.S. Army Center for Health Promotion and Preventive Medicine
- E. G. Daxon, Battelle, San Antonio
- T. T. Little, Los Alamos National Laboratory
- J. J. Whicker, Los Alamos National Laboratory
- Y. S. Cheng, Lovelace Respiratory Research Institute
- R. J. Traub, Battelle, Richland
- G. M. Lodde, Battelle Eastern Science and Technology Center
- F. Szrom, U.S. Army Center for Health Promotion and Preventive Medicine
- D. E. Bihl, Battelle, Richland
- K. L. Creek, Los Alamos National Laboratory
- C. B. McKee, Battelle Eastern Science and Technology Center

The independent peer reviewer science work group for this study was led by Dr. Roy Reuter and included Dr. Paul Baron, Dr. Wesley Bolch, Dr. John Doull, Dr. David Hoel, Dr. Anthony James, Dr. Morton Lippmann, Dr. Paul Strickland, Dr. Arthur Upton, and Dr. Wesley Van Pelt. The reviewers for the modeling methodology were Dr. Wesley Bolch, Dr. Keith Eckerman, Dr. Anthony James, and Dr. Otto Raabe.

Analysis of *Level II and Level III Inhalation and Ingestion Dose Methodology: Calculations and Results* (Attachment 4) was conducted by the following team:

- F. Szrom, U.S. Army Center for Health Promotion and Preventive Medicine
- E. G. Daxon, Battelle, San Antonio
- M. A. Parkhurst, Battelle, Richland
- G. A. Falo, U.S. Army Center for Health Promotion and Preventive Medicine
- J. W. Collins, U.S. Army Center for Health Promotion and Preventive Medicine

A very long list of people from many organizations supported the Capstone field tests and the Human Health Risk Assessment in one capacity or another and whose efforts are truly appreciated by the authors. The listings of organizations and individuals who provided the most support are acknowledged in the individual reports (Attachments 1 and 3).

Contents

Acknowledgments iii
Figures and Tables vii
Preface ix
Executive Summary xv

1.0 Introduction 1
    1.1 Purpose and Scope 4
    1.2 Approach 4
    1.3 Document Overview 7
    1.4 References 9

2.0 Characteristics of Depleted Uranium 11
    2.1 Uranium and Depleted Uranium Properties 12
    2.2 Characteristics of DU Oxides Generated by Armor Perforation . 13
    2.3 Health Risks of Depleted Uranium 17
        2.3.1 Chemical Health Risks 18
        2.3.2 Radiological Health Risks 19
        2.3.3 External Exposures 21
    2.4 References 21

3.0 Level I Exposures 23
    3.1 ODS Crewmember Exposure Incidents 23
    3.2 Modeling Parameters and Assumptions 24
        3.2.1 Inhalation Exposure Scenarios 25
        3.2.2 Physiological Parameters 26
    3.3 Chemical and Radiological Dose Models 27

    3.4 Aerosol Characteristics and Source-Term Data 27
    3.5 Capstone Dose and Risk Assessment Results 28
          3.5.1 Capstone HHRA Chemical Toxicity Results 29
          3.5.2 Capstone HHRA Radiological Dose and Risk Results ...31
          3.5.3 Multiple Perforations 35
    3.6 Comparison with Other Estimates 36
    3.7 References 40

4.0 Level II and Level III Exposures 43
    4.1 Methods 45
          4.1.1 Inhalation 46
          4.1.2 Ingestion 49
          4.1.3 Summary of Inhalation and Ingestion for Level II and
                Level III Exposures from the Capstone Data 53
    4.2 Summary of Level II and Level III Exposures 55
          4.2.1 Level II Estimates 55
          4.2.2 USACHPPM Level II Results of DU Munitions Cleanup ..57
          4.2.3 USACHPPM Level III Results 58
    4.3 Comparison with Other Estimates 59
    4.4 References 60

5.0 Depleted Uranium Chemical and Radiological Risk
in the Military Context 63
    5.1 System for Placing DU Radiological and Chemical Risks
         into Operational Context 64
          5.1.1 Radiological Health Risk 65
          5.1.2 Chemical Health Risk 67
    5.2 Operational Impact of the Estimated Exposures 68
    5.3 References 71

6.0 Putting the Results into Perspective 73
    6.1 Level I Exposures 74
          6.1.1 Crewmember Exposure Incidents 75
          6.1.2 Predicted DU Concentrations and Doses for
                Crewmembers and First Responders 75
    6.2 Level II and Level III Exposures 78
    6.3 Summary 78
    6.4 References 80

Abbreviations and Acronyms 81

Glossary 83

Figures and Tables

FIGURES

3.1 HHRA Median Peak Kidney Uranium Concentrations and 10th and 90th Percentiles for All Phases and Scenarios 31

3.2 HHRA Median Radiation Risks with 10th and 90th Percentile Bars from Perforation of the Vehicles 35

3.3 Estimates of Crewmember Committed Effective Dose Equivalent Probability Distribution for the Upper-Bound Calculation from USACHPPM (2000) 38

4.1 Probability Distribution of the Fraction of DU on the Hand Actually Ingested 51

TABLES

2.1 Composition of the Uranium Isotopes and Immediate Progeny in the U.S. DoD's DU 13

2.2 Renal Effects Groups—Chemical Risk to the Kidneys 19

3.1 Summary of Level I, In-Vehicle Personnel Exposure Scenarios ... 26

3.2 Summary of HHRA Median DU Intakes 28

3.3 Summary of HHRA Median Peak Kidney Uranium Concentrations 30

3.4 Summary of HHRA Median 50-yr Committed Effective Doses by Scenario 32

3.5 Summary of HHRA Median 50-yr Committed Equivalent Doses to the Lung by Scenario 33

3.6 Summary of HHRA Median Lifetime Risk of Fatal Cancer from DU Inhalation by Scenario 34

3.7 Comparison of Median Doses and Concentration Value 39

3.8 Comparison of Capstone HHRA Median and 90th Percentile 1-h Dose and Concentration Estimates with Worst-Case Estimates Made by The Royal Society 39

4.1 P-CI Summary Results for Level II Personnel Exposures In and Around Vehicles 47

4.2 CI-Array Summary Rate Results for Level II Area Monitors Inside Vehicles 48

4.3 Summary of IMBA-URAN Dose Conversion Factors Used 52

4.4 Summary of Ingestion Level II Monte Carlo Results Based on Mean Glove Contamination 52

4.5 Summary of Ingestion Level III Monte Carlo Results Based on Mean Glove Contamination 52

4.6 Approach to Estimating Potential Inhalation Doses 53

4.7 Summary of the Estimated E(50) and Peak Kidney Uranium Concentration Rates from Inhalation and Ingestion for Level II and Level III Personnel Exposures 54

4.8 Best-Estimates of Level II Exposures from Capstone Data and USACHPPM Data Assuming No Use of Personal Protection Equipment 55

4.9 Estimated Radiation Doses and Kidney Uranium Concentrations for Camp Doha Cleanup Personnel 57

4.10 Level III Exposure Estimates from USACHPPM (2000) 58

4.11 Exposure Estimates of Camp Doha Personnel Not Involved in Cleanup and Personnel Downwind of the Fire 59

4.12 Comparison of The Royal Society (2001, 2002), USACHPPM (2000), and the Capstone Study Estimated Radiation Dose Rates and Peak Kidney Uranium Concentrations for Level II Personnel 59

5.1 Comparison of RES Categories with "Peacetime" Standards or Guidelines 65

5.2 Summary of Radiological and Chemical Health Risk Categories ... 69

6.1 Predicted Median Peak Kidney Uranium Concentrations, Doses, and Risks to Vehicle Crewmembers 76

6.2 Predicted Median Peak Kidney Uranium Concentrations, Doses, and Risks to Vehicle First Responders 76

Preface

The U.S. Armed Forces first used a new kind of large-caliber (LC), anti-armor munition with a depleted uranium (DU) penetrator in the 1991 Gulf War, referred to as Operation Desert Storm (ODS). Use of this new munition helped to quickly end that war because of the effectiveness of the high-density munition to reach and perforate distant armored targets. During the course of ODS, misidentification of U.S. forces in distant vehicles led to a number of incidents in which U.S. armored vehicles were struck by the new LC-DU munitions. These incidents involved crews of six Abrams tanks and fourteen Bradley Fighting Vehicles (referred to as Bradley vehicles). One hundred four crewmembers survived. Some members of this population were struck by fragments from the DU penetrators that impacted their vehicles. They were treated for their injuries and most continue to be medically monitored. To date, no clinical symptoms of DU toxicity have been observed in this group.

Three of the Abrams tanks and most of the Bradley vehicles struck by the DU munitions had perforations of the crew compartments. When a DU munition penetrates armor, some of the penetrator metal erodes and forms a DU oxide powder. Crewmembers in these vehicles would have been exposed to DU oxide aerosols in addition to any DU fragments resulting from impact. This generated a great deal of concern, especially among the veterans of this conflict because of its low-level radioactivity. Although uranium in various forms, including DU oxides, is a much-studied material and its toxicity is relatively well known, the concentration of DU aerosols inside vehicles impacted and perforated by DU munitions had not been adequately studied to predict health effects from this exposure. This DU aerosol source-term information is critical for dose and risk assessment.

Since the inception of the DU munitions in the early 1970s, the U.S. Department of Defense (DoD) has periodically conducted evaluations of the radiological safety and environmental impacts of the military use of depleted uranium. Results of these studies are summarized in reports by the Office of the Special Assistant to the Deputy Secretary of Defense for Gulf War Illnesses, known as OSAGWI, (1998, 2000) and the U.S. Army Center for Health Promotion and Preventive Medicine, abbreviated as USACHPPM, (2000). The USACHPPM report reviewed the available data related to personnel exposures to DU ammunition and DU oxides created by impact with armor and from munition fires. USACHPPM determined that the personnel exposure data were relatively robust except for the data required to estimate exposures for those personnel in, on, or near (within 50 meters) an armored vehicle when it was perforated by a DU munition or DU armor perforated by any munition. To fill this gap in the database, the U.S. Army and OSAGWI sponsored the Capstone DU Aerosol Characterization and Risk Assessment Program. The purpose of the Capstone effort was to provide a peer reviewed, rigorous scientific estimate of the health risks to military personnel in and around armored vehicles perforated by a large caliber DU munition.

The Capstone Program had two major components. The first component required the generation and characterization of DU aerosols created by the perforation of an Abrams tank (Figure P.1) and a Bradley Fighting Vehicle (Bradley vehicle) (Figure P.2) with a large-caliber DU penetrator. The tests

Figure P.1.
Abrams Main Battle Tank

Figure P.2.
Bradley Fighting Vehicle

that formed this component required the design and fabrication of aerosol sampling systems rugged enough to survive perforation of the crew compartment and complex enough to collect DU particles in the air of the crew compartment as a function of time. The aerosol particles collected inside the crew compartment (a combination of DU oxide powder and particles from the perforated armor) were characterized to quantify their DU content and determine other properties that affect their behavior if inhaled into the body. This project provided the robust data required to estimate the amount of DU internalized through inhalation and ingestion so that the radiological dose and the chemical concentrations in the body could be calculated.

The second component of the Capstone Program was the Human Health Risk Assessment (HHRA) that used internationally recognized models to estimate the radiological dose and chemical concentrations in the body and translate these values into estimates of risk. Scenarios were developed based on the amount of time crewmembers remained in the perforated vehicle (stay-times). These included the "most likely" stay-times of 1 and 5 minutes (min) that best represented the ODS experience, especially with the Abrams tanks, and "upper bound" stay-times of 1 and 2 hour (h). Stay-time is an important factor in the analysis because it is related to the amount of DU that was internalized. The doses were not linear with time, but longer stay-times resulted in higher doses. Additionally, a scenario was developed to represent first responders, those military personnel first on the scene (who were not in the vehicle) who help with vehicle and equipment evacuation.

The DU aerosol concentrations, stay-times, and breathing rates were key parameters used to estimate the amount of DU these personnel could have internalized. Advanced modeling techniques and the Capstone-determined particle properties allowed an estimate of the DU intakes, DU concentrations in the tissues and organs in the body, and the associated radiological doses. A detailed review of the extensive body of literature on uranium health effects was used to determine the relationship between the amount of DU in the body and the resulting chemical and radiological health risks.

Placing these risks into perspective is complicated because of the nature of the health effects (kidney effects, potential increase in cancer) and the standards of "acceptable risk" usually used as a basis for comparison. U.S. "peacetime" occupational guidelines and limits for chemical concentrations and radiation dose were the starting points for evaluating levels of exposure protective of health. However, these standards were not designed to be applied in the potentially lethal environments that characterize military combat operations.

The DoD recognized this operational reality and developed a system of radiation controls that was based on U.S. radiation standards for normal occupational settings and for emergent situations where the risks of inaction due to radiation exposure are far greater than the risks of the exposure. This combination of peacetime occupational standards and emergency standards provide the ability to match the radiation risk with a military operational risk in a manner that allows risk management decisions to mitigate the total risk (operational and radiation) to the Soldier. The Institute of Medicine reviewed this system and concurred with the approach (Mettler 1999), and the HHRA applied this approach in an evaluation of military risk management.

Because a DoD system for DU/uranium chemical toxicity similar to the system for radiation exposure had not been developed, the HHRA developed a system to quantify chemical risks. This system was based on human data that related the amount of uranium in the kidney to harmful effects to the kidney. The current *de facto* guideline for chemical toxicity is a kidney uranium concentration of 3 micrograms (μg; one millionth of a gram) of uranium per gram of kidney. The chemical effects of depleted uranium and natural uranium are the same. In fact the only difference is that depleted uranium is 40% less radioactive than the small amounts of natural uranium each of us eats, drinks, and breathes on a daily basis.

The HHRA analysis predicted exposures for crewmembers in an Abrams tank and a Bradley vehicle at the time of perforation and for first responders who entered the perforated vehicle shortly after the event. These groups of crewmembers and first responders are denoted as being within the Level I category of potential exposure. Individuals within this group are most likely to be exposed to the highest levels of DU in a combat situation. General conclusions of the HHRA for Level I exposures are the following:

- For all vehicle configurations and modeled exposure times, except for the *unventilated* Abrams tank perforated through DU armor, predicted radiation doses were within U.S. (routine) occupational limits. For the *unventilated* Abrams tank perforated through DU armor, short exposures (about 1 min) were within routine occupational limits, and exposures up to 2 h were within the emergency or planned special exposure limits. ***For all vehicle configurations and exposure times modeled (up to 2 h), predicted radiation doses are not likely to cause adverse health effects.***

- For all vehicle configurations and exposure times, except for the *unventilated* Abrams tank perforated through conventional armor, ***predicted uranium concentrations in the kidney are not likely to cause adverse***

chemically-induced health effects. In the case of the *unventilated* Abrams tank perforated through conventional armor, the potential exists for short-term adverse kidney effects for exposures 5 min or longer.

- Although there is no compelling evidence from human epidemiologic studies that internal uranium exposure is associated with an increased cancer risk, generic cancer risk factors and conservative modeling assumptions were used to estimate cancer risks from the radiological doses calculated. The increased lifetime risks of fatal cancer from inhaling DU aerosols ranged from a low of 0.005% for a 1 min exposure in a ventilated Abrams tank perforated through DU armor to a high of 0.44% for a 2-h exposure in an unventilated Abrams tank perforated through DU armor. The Bradley vehicle and Abrams tank perforated through conventional armor fell between these extremes. For perspective of the meaning of these risk levels, the increase above U.S. baselines in lung cancer rates due to smoking in U.S. males is 7.35%, and the natural lifetime risk of fatal cancer among all U.S. males is 23.6% (Ries et al. 2003) while the modeled exposure to DU would increase this risk by considerably less than 1%.

Specific DU intakes and associated chemical and radiological doses and risks are listed in Section S.3 by vehicle type and length of exposure and are summarized in Section S.6.

This report also summarizes the DU intakes, uranium chemical and radiological doses, and risk estimates for other personnel who came in contact with DU without personal protective equipment, which prevents inhalation of DU particulates. This includes those whose jobs required them to work inside vehicles with DU residues or to clean up DU residues (denoted as Level II exposures) and those personnel who came into incidental contact with DU (denoted as Level III) either through brief entry into an uncleaned perforated vehicle or through other means such as being downwind of fires involving DU. In general, the radiological doses and kidney uranium concentrations for these groups are well below occupational exposure limits with the possible exception of unprotected Level II personnel who spend long periods of time working in uncleaned vehicles. These exposures can be readily mitigated with simple precautions including the use of respiratory protection to avoid inhaling the residues, and the use of gloves and hand washing to avoid ingesting them.

By default, Level I personnel are in combat. The HHRA showed that the risks of DU are low enough that only those mitigation actions that do not increase the already grave operational risk should be undertaken. The most effective and simplest procedure is to turn on ventilation systems if they are operative and are not already operating. If ventilation systems operate as intended, exposures are estimated to be below both the radiological and chemical occupational standards.

As with many of the other possible hazards on the battlefield, the DU exposure risk levels for Level I and Level II personnel are in the range that hazard awareness training constitutes an effective risk mitigation measure. Appropriate post-operation medical follow-up (biomonitoring) should still be conducted to assess individual exposures. Awareness training is also appropriate for Level III personnel, even though the risks of DU exposure are greatly decreased.

Executive Summary

This Summary Report and its attachments[a] document the results of the Capstone DU Aerosol Characterization and Risk Assessment Program, which focused on exposures to DU aerosols inside perforated armored vehicles. Information on exposures outside a perforated vehicle and from exposure to DU aerosols from ammunition fires is also presented within this Summary. The following paragraphs describe this Summary Report, the four reports resulting from the Capstone program, and the letters identifying overall test objectives on which the program was based are provided.

Summary Report

This report summarizes the results from the study to characterize DU aerosols generated during perforation of an Abrams tank and a Bradley vehicle and from the HHRA. Discussions of predicted DU exposures and health effects are subdivided by the probable exposure levels. Three levels (I, II, and III) define exposure groups by time and duration of exposure occurrence as well as activity level. Level I exposures pertain to personnel in, on, or near the combat vehicle at the time of perforation as well as those personnel who respond immediately, post-perforation to render aid—first responders. Individuals within the Level I category and inside the vehicle at the time of perforation would receive the highest DU exposures. Level I first responders would receive lower exposures. Level II exposures typically involve DoD personnel who work in and around vehicles containing DU fragments and

---

[a]Attachments 1 through 5 available on CD and Attachment 3 available in hard copy through Battelle Press. Attachments 1 and 2 also available through NTIS.

particles, while Level III exposures pertain to personnel whose exposures were incidental, brief, and often occurred several hours after perforation.

This report also summarizes the Level I exposure results documented in Attachments 1 through 3 (discussed below). Data provided by the USACH-PPM (2000) report are used to support the analysis of Level II and Level III exposures, revising or updating them where possible with data from the Capstone program as discussed in Attachment 4.

DU Aerosol Generation and Characterization Report (Attachments 1 and 2)

The generation and characterization of DU aerosols in perforated Abrams tanks and Bradley vehicles are documented in the report entitled *Capstone Depleted Uranium Aerosols: Generation and Characterization, Volumes 1 and 2* (Parkhurst et al. 2004). Volume 1 of that study, hereafter referred to as the Capstone DU Aerosol Study or the Capstone Study, is Attachment 1 to this Summary Report. Volume 2 of the Capstone Study report (Attachment 2 to this Summary Report) contains seven appendices with detailed characterization data.

The field tests focused primarily on collecting and characterizing DU aerosols generated inside the vehicles for use in evaluating intake from inhalation and calculating the resulting doses. Sampling was also conducted to support an analysis of DU ingestion (hand-to-mouth transfer).

Human Health Risk Assessment, Level I Inhalation Exposures (Attachment 3)

The evaluation of health effects of the DU aerosols measured during the Capstone field tests is provided in *Human Health Risk Assessment of Capstone Depleted Uranium Aerosols* (Guilmette et al. 2004). That report, referred to as the HHRA is Attachment 3 of this Summary Report. The assessment was limited to the evaluation of adverse health effects from inhalation for Level I category DU exposures, specifically to crewmembers in the vehicle at the time of perforation. Only DU aerosol inhalation was modeled. The risk assessment used scientific guidance and models for breathing factors, tissue/organ weighting factors, and biokinetic and internal radiation dosimetry. The models were used to calculate DU intakes, DU concentrations in tissues and organs, and radiological doses to tissues and organs. These models used both a conventional approach to statistical evaluation and a Bayesian statistical approach to calculate intakes, organ concentrations, and doses, and to assess variability.

Because uranium is radioactive, the possibility of radiogenic cancer induction was evaluated with a focus on lung cancer. Chemical toxicity was evaluated based on reviews of occupational studies of uranium workers and was augmented with results of animal studies. The kidney is considered the primary target organ for uranium chemical toxicity and was the main focus of the chemical toxicity assessment, although toxicity to other organs was also addressed. Renal Effects Groups (REGs) were developed and used to predict kidney effects from DU intake. (Refer to Section 2.3.1 for a discussion of the REGs.)

Level II and Level III Inhalation and Ingestion Exposures (Attachment 4)

Aerosol samples were collected during the Capstone Study and support the analysis of exposure to residual aerosols, resuspended aerosols, and hand-to-mouth ingestion of DU. In this attachment, the processes used to estimate inhalation and ingestion doses for Level II and Level III exposures are discussed and the results are presented. These results are summarized in Chapter 4 of this Summary Report.

Data Quality Objectives (DQOs) for the Capstone Study and the HHRA (Attachment 5)

An Army steering committee (the Depleted Uranium Research—Integrated Process Team [DUR-IPT]) guided the overall test objectives of the Capstone Study and the HHRA and test implementation of the Capstone Study. The U.S. Army Medical Command (MEDCOM) developed a set of DQOs for the specific information to be derived from this testing program. An independent nine-member peer review panel provided technical feedback on draft DQOs. This feedback was used in developing the final DQOs. The letters outlining the DQOs for the two phases of the program are provided in this attachment.

S.1 DU AEROSOL CHARACTERIZATION

The field tests used to collect aerosols for characterization were conducted in four phases. In the first three phases, structurally intact target vehicles (Abrams tank and Bradley vehicle ballistic hulls and turrets [BHTs]) were used. The BHTs were stripped of flammable equipment and instrumentation that were not needed for the test and could have compromised sample collection. These vehicles did not have any ventilation systems, and as a result, represent a reasonable model for upper bounds of aerosol concentrations expected under operational conditions. In the fourth phase, the target vehicle was an operating, fully functional Abrams tank, and its environmental control/nuclear, biological, chemical (EC/NBC) overpressure ventilation system was operating at the time of impact and during and after perforation. This system effectively reduced the aerosol concentrations and associated doses. The Halon fire-suppression systems were not activated in the tests.

The four test phases and the target vehicle configurations were as follows:

- Phase I—Abrams tank (BHT) perforated through conventional (non-DU) armor, no ventilation
- Phase II—Bradley vehicle (BHT) perforated through conventional armor, no ventilation
- Phase III—Abrams tank (BHT) perforated through DU armor, no ventilation
- Phase IV—Abrams tank perforated through DU armor, EC/NBC ventilation system operating.

Aerosol samplers were selected to provide the types of data required for the subsequent dose and health risk assessments. These samplers included filter cassettes, 8-stage cascade impactors, a 5-stage cyclone separator, and a moving filter sampler. Limited sampling outside the vehicle was also conducted using high-volume air samplers in the first few experiments and cascade impactors for the remaining experiments. Wipe survey samples provided additional data for evaluating DU ingestion.

Generalizations related to the physical and chemical characteristics of the DU aerosols as collected inside the target vehicles are summarized briefly below and discussed in detail in Chapter 2 of this Summary Report.

- The uranium mass, expressed as a percentage of the total mass of aerosol collected, varied by target vehicle and by particle size. The percentage in the collected aerosol that was uranium ranged by vehicle as follows:

  —38 to 72% in the Abrams tank perforated through conventional armor

  —60 to 72% in the Abrams tank perforated through DU armor

  —18 to 29% in the Bradley vehicle perforated through conventional armor.

  The remainder of the aerosol consisted of oxygen, aluminum, iron, and very small or trace amounts of other metals including titanium (alloyed with DU in the penetrator), zinc, and copper.

- DU aerosol concentrations decreased rapidly as a function of time as settling occurred. Within 30 min, DU aerosol concentrations were typically about 100 times lower than the concentrations measured just after perforation.

- Particle-size distributions changed as a function of time as settling occurred. The larger particles tended to settle out more quickly than did the smaller particles throughout the settling process. Particle shapes were relatively heterogeneous.

- The predominant uranium oxide phase was U_3O_8/UO_3. This fraction increased as particle size decreased, while the percentage of U_4O_9, which was highest in the large particles, decreased as particle size decreased. A small amount of $UO_3 \cdot 2H_2O$ (schoepite) was detected in several samples.

- The dissolution rates in simulated lung fluid of the DU aerosols varied by sample. Between 58 and 99% of the sample material dissolved slowly (half-time of >100 days). A range between 1 and 36% of the sample material dissolved quickly (half-time of <10 days). Several samples also contained an intermediate (half-time between 10 and 100 days) component. The results most closely resembled the Type M (a moderate rate of dissolution) absorption behavior as defined in the International Commission on Radiological Protection (ICRP) Publication 66 (1994).

S.2 LEVEL I HUMAN HEALTH RISK ASSESSMENT

The HHRA evaluated inhalation exposures and associated DU tissue/organ uranium concentrations, radiation doses, and health risks for Level I exposures to personnel (crewmembers inside the vehicle at the time of perforation and first responders who entered the vehicle within a few minutes after perforation). The analysis used specific exposure times and durations to develop five scenarios that were used to evaluate doses and risks. These exposure durations ranged from 1 min to 5 min to 2 h to cover most likely and upper bound exposures. In the three incidents that occurred during ODS, surviving Abrams tank crewmembers who were in the tank when the crew compartment was perforated exited the vehicle within 2 min. The Halon fire-suppression systems activated in all three incidents. The risk assessment was extended to 2 h to address the increase in doses that could result from possible extended exposure in future combat. Breathing rates representative of light and heavy exercise were used (see Table 3.1)

The target vehicles used in all but Phase IV tests contained no ventilation systems. Aerosol concentrations would be higher in these vehicles than in vehicles with operating ventilation systems. Therefore, applying results from unventilated vehicles to circumstances in which one or more ventilating system was operating would overestimate exposures. No specific information was gathered about the use of ventilation systems at the time of or immediately following perforation during ODS. However, standard practice would have been to have an onboard ventilation system operating. Consequently, actual Abrams tank experience is likely to have been more similar to risk assessment results for the Capstone Phase IV firing test, in which the EC/NBC ventilation system was operating. In addition, unlike the three Abrams tank incidents during ODS in which the crew compartment was perforated and the Halon systems were activated, the onboard Halon systems were either not present or were not activated during the Capstone tests. Personal respirators were not used in these vehicles.

The predicted median intakes of DU from the aerosols for the most likely scenarios are listed for each vehicle configuration in Table S.1. The intakes range from a low of 10 milligram (mg) for a 1 min exposure in the ventilated Abrams tank with DU armor to a high of 710 mg for a 5 min exposure in an unventilated Abrams tank with DU armor. These data were used as input to the models calculating DU concentrations to tissues and organs and radiological doses.

Table S.1. Summary of Median Intakes for the Most Likely Scenarios

Most Likely Scenarios	Intake (mg)			
	Abrams Tank: Conventional Armor, No Ventilation	Abrams Tank: DU Armor, No Ventilation	Abrams Tank: DU Armor and EC/NBC Operating	Bradley Vehicle: Conventional Armor, No Ventilation
A - Crew, exit in 1 min	280	250	10	83
B - Crew, exit in 5 min	590	710	43	220
E - First responders, entry at 5 min, exits 10 min later	160	200	27	99

Peak kidney uranium concentrations, doses, and risks for Level I exposures are summarized in Chapter 3 of this Summary Report. Results of the most realistic estimates of dose and risk for short-term exposure, as determined from the Capstone data, are discussed in the following sections. Discussions of upper bound results from extended occupancy (up to 2 h, Scenarios C and D) of perforated vehicles (assuming presence during perforation) are provided in Chapter 3. Most of the radiological data are presented using the traditional units of nanocuries (nCi; 0.38 nCi/mg DU) and rem (100 rem/1 sievert [Sv]).

Chemical Toxicity Assessment

Depleted uranium is a form of uranium that is considered to be "weakly" radioactive because, through processing, its radioactivity is 40% lower than that of natural uranium. It is also a heavy metal. Uranium is ubiquitous in the environment in the air we breathe, the food we eat, and the water we drink. Levels of uranium in drinking water vary widely, and trace amounts are found in foods. The human body naturally contains some uranium, primarily from water intake, without apparent adverse health effect. Historically considered a "feeble poison" when ingested, it can cause heavy metal toxicity if the intake is at high levels. Uranium has been demonstrated to affect several organ systems after exposure to sufficiently high concentrations; the most important of these organs are the kidneys and the lungs.

DU concentrations from Level I intakes were evaluated in the HHRA. Peak kidney uranium concentrations for DU occur about one day after exposure. These peak kidney uranium concentrations were used to calculate the median, standard deviation, and 10th and 90th percentiles (provided in Attachment 3, Section 6.4) for each phase and scenario. Results from the Level I analysis are shown in Table S.2, which summarizes the median peak kidney uranium concentrations for each tested vehicle configuration for the two "most likely" crewmember scenarios and the "most likely" first responder scenario.

Levels near or above the *de facto* occupational guideline of 3 µg U/g kidney were predicted in these unventilated vehicles in which crewmembers exited about 5 min after perforation. The highest kidney uranium concentrations occurred for exposures in the Abrams tank perforated through conventional armor (Phase I). Comparison of these peak kidney uranium concentrations with documented human cases predicts that individuals exposed at these levels are not likely to become ill with the possible exception of those exposed for 5 min in the Abrams tank with conventional armor.

Predicted peak kidney uranium concentrations for a crewmember exposed in an Abrams tank (perforated through DU armor) in which the EC/NBC

Table S.2. Summary of Median Peak Kidney Uranium Concentrations for Most Likely Scenarios

Most Likely Scenarios	Peak Kidney Uranium Concentrations (µg U/g Kidney)			
	Abrams Tank: Conventional Armor, No Ventilation	Abrams Tank: DU Armor, No Ventilation	Abrams Tank: DU Armor and EC/NBC Operating	Bradley Vehicle: Conventional Armor, No Ventilation
A - Crew, exit in 1 min	3.0	1.1	0.05	1.0
B - Crew, exit in 5 min	6.4	2.6	0.23	2.9
E - First responders, entry at 5 min, exits 10 min later	1.5	0.7	0.14	1.4

ventilation system was operating were more than 10 times less than the uranium concentrations for exposures in an Abrams tank (perforated through DU armor) without an operating ventilation system.

Radiological Dose Assessment

The primary radiological health risk for inhaled DU is a potential for an increased risk of developing cancer. There is, however, no compelling evidence from human epidemiologic studies that uranium internal exposures are associated with an increased cancer risk. As a result, instead of using cancer risk factors specifically related to DU, the analysis of the risk of developing cancer is based on the general theoretical radiation risks of alpha emitters, of which uranium is one. Biokinetic model calculations indicate that the risk of internally deposited DU is primarily to the lung.

In general, the risk of developing cancer is related to the dose. *Median 50-yr committed effective doses,* E(50), for the three "most likely" scenarios are presented in Table S.3. The E(50)s for the "most likely" exposures in the Abrams/conventional armor and the Bradley vehicle were below 5 rem, the U.S. limit for routine occupational exposures. The E(50) for Scenario B in the unventilated Abrams tank perforated through DU armor was predicted to be 6 rem. Because the full E(50) dose accrues over 50 years instead of a single year, the E(50) values do not directly correlate with annual doses. The true annual dose is much less than the 50-yr committed dose, both for individual organs and for the effective dose. Even so, these doses are well below the U.S. Nuclear Regulatory Commission's (NRC's) annual dose limit of 10 rem/yr for occupational workers with a planned special exposure (e.g., protecting critical property during an emergency).

Unlike the results of the toxicology analysis, the radiological doses were highest from the Abrams tank perforated through DU armor (Phase III) because these aerosols were slightly more insoluble than the aerosols collected from Phase I. Doses to a crewmember in an Abrams tank with its

Table S.3. Summary of the Median Committed Effective Doses for Most Likely Scenarios

Most Likely Scenarios	E(50) (rem)			
	Abrams Tank: Conventional Armor, No Ventilation	Abrams Tank: DU Armor, No Ventilation	Abrams Tank: DU Armor and EC/NBC Operating	Bradley Vehicle: Conventional Armor, No Ventilation
A - Crew, exit in 1 min	2.0	2.2	0.09	0.6
B - Crew, exit in 5 min	3.7	6.0	0.44	1.7
E - First responders, entry at 5 min, exits 10 min later	0.9	1.9	0.41	0.9

EC/NBC system operating when perforated through DU armor were 2 to 22 times less than the levels in a perforated Abrams tank without an operating ventilation system. Radiation doses at the levels predicted are not expected to cause adverse health effects.

Fifty year committed equivalent doses, $H_T(50)$, to organs were also calculated for the lung, bone surface, kidney, red marrow, and liver. The doses to the lung were higher by at least a factor of 10 than the doses to these tissues/organs. The predicted *median* $H_{Lung}(50)$s calculated for the lung are summarized in Table S.4 for each of the three most likely scenarios. Exposures that produce $H_{Lung}(50)$s at the levels predicted are not expected to cause adverse health effects. Organ doses to a crewmember in an Abrams tank with its EC/NBC ventilation system operating when perforated through DU armor were at least 2 and up to 22 times less than the levels in a perforated Abrams tank in which the ventilation system was not operating.

Even though DU has not been shown to cause cancer, the risk of fatal cancer from inhalation of DU aerosols was calculated from generic risk coefficients using a summed organ risk approach. The calculations used the conservative Linear, No-Threshold model of effect that may overestimate risks at the levels predicted in this study and are, therefore, thought to be conservatively protective of health. Table S.5 summarizes the median estimated increased

Table S.4. Summary of the Median Committed Equivalent Lung Doses for Most Likely Scenarios

Most Likely Scenarios	$H_{Lung}(50)$, rem			
	Abrams Tank: Conventional Armor, No Ventilation	Abrams Tank: DU Armor, No Ventilation	Abrams Tank: DU Armor and EC/NBC Operating	Bradley Vehicle: Conventional Armor, No Ventilation
A - Crew, exit in 1 min	14	18	0.7	5.2
B - Crew, exit in 5 min	32	44	3.3	14
E - First responders, entry at 5 min, exits 10 min later	8.8	14	3.1	6.7

Table S.5. Summary of Median Lifetime Risk Increase of Fatal Cancer from DU Inhalation for Most Likely Scenarios

Most Likely Scenarios	Lifetime Risk Increase of Fatal Cancer (%)			
	Abrams Tank: Conventional Armor, No Ventilation	Abrams Tank: DU Armor, No Ventilation	Abrams Tank: DU Armor and EC/NBC Operating	Bradley Vehicle: Conventional Armor, No Ventilation
A - Crew, exit in 1 min	0.11	0.12	0.005	0.03
B - Crew, exit in 5 min	0.20	0.32	0.025	0.10
E - First responders, entry at 5 min, exits 10 min later	0.05	0.10	0.023	0.05

lifetime cancer mortality from inhaled, deposited DU aerosols for the most likely scenarios evaluated.

The median probability of lifetime cancer mortality from DU aerosol inhalation within the "most likely" exposure levels ranged from 0.005 to 0.32%. A risk of 0.32% implies that, on average, the chance over an individual's lifetime of dying of cancer from this exposure is 0.32% greater than the chance of that same individual dying from the natural or background rate of cancer mortality, which is 23.6% for U.S. males (Ries et al. 2003). Risks to a crewmember in an Abrams tank (perforated through DU armor) in which the EC/NBC ventilation system was operating were a factor of at least 2 and up to 22 less than the risks in an Abrams tank (perforated through DU armor) without an operating ventilation system.

S.3 LEVEL II AND LEVEL III INHALATION AND INGESTION EXPOSURES

Potential inhalation exposures from resuspended DU aerosols were evaluated for personnel entering the vehicle 2 h or more after perforation. This evaluation was conducted using personal samplers worn by the field recovery crews and area samplers within the perforated vehicles. Lower bound inhalation intake and dose rates were estimated using data from breathing zone personal samplers. Upper bound inhalation intake and dose rates were estimated using the area sampler measurements.

Hand-to-mouth ingestion intake and dose rates from DU deposited on surfaces were estimated from measurements of DU contamination on cotton gloves worn by members of the sample recovery team. These data were supplemented with wipe data from surface deposition of aerosols (discussed in Attachment 4, Section 3.2).

Table S.6 summarizes the intake and dose rates estimated for Level II and Level III personnel from inhalation and ingestion exposures. The results presented are for a single acute intake and assume that personal protective equipment (PPE) is not used and that decontamination has not taken place prior to the individual's entry into the vehicle. Decontamination procedures would reduce the amount of DU residues available for resuspension and ingestion.

For inhalation exposures, the table provides two estimates for DU intake, E(50), and peak kidney uranium concentration rates. The lower bound inhalation estimates are based on the mean values of breathing zone monitor samplers worn by sample recovery personnel for all vehicle types. The upper

Table S.6. Estimated Intake, E(50), and Peak Kidney Uranium Concentration Rates from Inhalation and Ingestion for Level II and Level III Personnel Exposures

Parameter as a Rate	Inhalation (Levels II and III)		Ingestion	
	Lower Bound Mean	Upper Bound Mean	Level II	Level III
DU Intake (mg/h)	0.447	14.5	10.6	1.78
E(50) (rem/h)	1.97E-03	7.80E-02	7.07E-04	1.20E-04
Peak Kidney U Conc. (µg U/g kidney-h)	2.89E-03	1.43E-01	7.67E-02	1.30E-02

bound estimates are based on area monitors that operated inside the vehicle during recovery operations following two Phase-I shots. The inhalation intake and dose rates are based on sampling during Level II-like activities. As a result, applying the results to Level III exposure scenarios will probably overestimate the Level III intakes and doses, whose exposures are characterized as incidental.

For ingestion exposures, the table provides two estimates for DU intake, E(50), and peak kidney uranium concentration rates, one for Level II personnel and one for Level III personnel. The results are based on the glove contamination data, which could be categorized as Level II-like or Level III-like activities. Ingestion intake and dose estimates for a single acute exposure can be obtained by multiplying the appropriate (Level II or Level III) rate by the time an individual was in contact with surfaces contaminated with DU residues. If vehicle wipe-test data are available for a suspected ingestion exposure, a method described in Attachment 4 (Section 3.2.2) can be used to estimate an intake and dose from the actual data.

The results presented above are for a single acute exposure. If multiple exposure duration times occur, the peak kidney uranium concentration rate cannot simply be multiplied by the total time because uranium clearance from the kidney will occur between the exposures. Therefore, with multiple exposures, the intervals of exposure as well as the time between the exposure intervals are necessary to estimate the peak kidney uranium concentration. The E(50) for multiple exposure duration scenarios can be calculated by multiplying the E(50) rate by the total time because E(50) is cumulative.

S.4 MILITARY MANAGEMENT OF DU EXPOSURES

The level of risk in a military operation ranges from those that are comparable to peacetime occupational environments to those that are orders of magnitude higher. In a peacetime environment, the risk associated with the use of a protective procedure is usually a matter of cost. In a military combat environment, the application of a protective measure may incur more risk than is averted. For example, donning chemical protective gear in a hot climate increases the risk of heat related injuries and reduces the Soldier's effectiveness in safely performing the job. Reduced visibility and reduced dexterity exacerbate the difficulty and risk associated with any task. The commander of military operations needs a system to balance immediate risk (morbidity and mortality from combat, operational accidents, and diseases) with risks from chemical or radiological exposure that may or may not manifest immediate or delayed adverse health affects. A process for doing this is discussed in Chapter 5.

Three conclusions related to military management of DU exposures can be drawn from the analysis of the Level I exposure risks compared to other combat risks.

- First, the chemical and radiation risks are low enough that normal combat crew-response drills should not be altered because of the presence of DU.

- Second, as with many of the hazards on the battlefield, the risk levels for crewmembers are high enough that hazard awareness training should be conducted and appropriate medical follow-up (biomonitoring) should be performed post operation to assess individual exposures (DoD 2004).

- Third, first responder risks are below those associated with "peacetime" annual occupational exposure limits.

S.5 CONCLUSIONS

The robust data from the Capstone DU Aerosol Study provide a sound basis for assessing the inhalation of DU aerosols and the dose and risk to personnel in combat vehicles at the time of perforation and to those entering immediately after perforation. The HHRA provided a technically sound process for evaluating chemical and radiological doses and risks from Level I DU aerosol exposure using a combination of innovative and well-accepted models. The Capstone Study also provided data useful in estimating Level II and Level III inhalation and ingestion dose rates from exposures to resuspended and deposited material.

The chemical and radiological doses and risks to human health of inhaling DU aerosols in a perforated vehicle are relatively low when compared with many other combat risks. In addition to the possibility of combat injuries, hazardous materials including different heavy metals, chemicals, and other agents may be released onto the battlefield and into the environment.

The most important factor for reducing exposure and dose immediately after vehicle perforation is the use of onboard vehicle ventilation, especially if it is not safe to exit the vehicle. Even though the doses and risks from inhaled DU based on the Capstone data are low, ventilation systems operating during or turned on as soon as possible after a DU perforation can significantly reduce the intake and its associated exposure received by the crewmembers.

The following are the major conclusions and practical generalizations drawn from this dose and risk assessment.

Predicted Doses, Risks, and Human Health Effects

- In Abrams tanks with one or more ventilation systems operating, the levels of predicted chemical toxicity and radiological risks for the DU exposure scenarios are below routine occupational limits. Personnel are not likely to develop adverse health effects as a result of exposure at these levels.

- In unventilated Abrams tanks perforated through DU armor, it is unlikely that adverse health effects would result from the most likely exposure scenarios of 5 min or less.

- In unventilated Abrams tanks perforated through conventional armor, no adverse health effects are predicted for a 1-min exposure. The potential exists for temporary kidney effects for exposures of about 5 min and longer.

- No adverse health effects are expected from exposures in unventilated Bradley vehicles for any of the time periods modeled (up to 2 h).

- In unventilated Abrams tanks, most of the intake would occur within the first 5 to 10 min after perforation. This intake would be significantly reduced by activation of the ventilation systems during this timeframe. With an operating ventilation system, the dose and, therefore, the risks would be reduced by about an order of magnitude over the risk incurred in vehicles with no ventilation systems operating during or after perforation. Dose and risk reductions would also be expected in a Bradley vehicle with an operating ventilation system.

- Although shot-to-shot variations in aerosol generation occur in combat situations, the redundancies in the study were sufficient to ensure reasonable certainties in the data and in the conclusions drawn from the data.

Other Considerations

- The veterans who sustained DU fragment wounds would have also inhaled DU aerosols as a result of armor perforations. Because of the DU fragments, their total intake of DU would be at least as large, and most likely larger, than the modeled Level I intakes, which only considered inhalation exposures. After more than a decade of medical surveillance of these 1991 Gulf War survivors of DU-related injuries, no adverse toxicological effects related to the presence of DU have been identified (McDiarmid et al. 2004). This supports the conclusion of this report that no adverse health effects, especially to the kidney, would be expected for the most likely exposures presented.

- Although radiation risks are predicted to be low, counseling of affected personnel and their family members is suggested because of the perceived radiological risks associated with exposure to DU. It would be helpful if the counseling included a discussion of other battlefield hazards and associated risks.

- Because of differences in individual exposures for a given crewmember in a perforated vehicle, DU urine bioassays are needed to establish individual dose estimates.

- Application of this study's data to Operation Iraqi Freedom or other combat situations is valid only for large-caliber DU munitions although many of the same aerosol characteristics, but not the concentrations, may be similar to exposures generated by smaller caliber DU munitions.

S.6 REFERENCES

Guilmette RA, MA Parkhurst, G Miller, FF Hahn, LE Roszell, EG Daxon, TT Little, JJ Whicker, YS Cheng, RJ Traub, GM Lodde, F Szrom, DE Bihl, KL Creek, and CB McKee. 2004. *Human Health Risk Assessment of Capstone Depleted Uranium Aerosols*. PNWD-3442, prepared for the US Army by Battelle under Chemical and Biological Defense Information Analysis Center Task 241, DO 0189, Aberdeen, Maryland.

International Commission on Radiological Protection (ICRP). 1994. *Human Respiratory Tract Model for Radiological Protection*. ICRP Publication 66, Pergamon Press, Oxford, United Kingdom.

McDiarmid MA, S Engelhardt, M Oliver, P Gucer, PD Wilson, R Kane, M Kabat, B Kaup, L Anderson, D Hoover, L Brown, B Handwerger, R Albertini, D Jacobson-Kram, C Thorne, and K Squibb. 2004. "Health Effects of Depleted Uranium on Exposed Gulf War Veterans: A 10-Year Follow-Up." *J. Toxicol. Envir. Health*, 67:277-296.

Mettler FA. Jr., Committee Chairman. 1999. *Potential Radiation Exposure in Military Operations: Protecting the Soldier Before, During, and After*. Committee on Battlefield Radiation Exposure Criteria, Institute of Medicine, National Academy Press, Washington, DC.

Office of the Special Assistant to the Deputy Secretary of Defense for Gulf War Illnesses (OSAGWI). 1998. *Exposure Investigation Report, Depleted Uranium in the Gulf*. Online report available at URL: www.gulflink.osd.mil in the Environmental Exposure Reports Section.

Office of the Special Assistant to the Deputy Secretary of Defense for Gulf War Illnesses (OSAGWI). 2000. *Exposure Investigation Report, Depleted Uranium in the Gulf (II)*. Online report available at URL: www.gulflink.osd.mil in the Environmental Exposure Reports Section.

Parkhurst MA, F Szrom, RA Guilmette, TD Holmes, YS Cheng, JL Kenoyer, JW Collins, TE Sanderson, RW Fliszar, K Gold, JC Beckman, and JA Long. 2004. *Capstone Depleted Uranium Aerosols: Generation and Characterization, Volumes 1 and 2*. PNNL-14168, prepared for the U.S. Army by Pacific Northwest National Laboratory, Richland, Washington.

Ries LAG, MP Eisner, CL Kosary, BF Hankey, BA Miller, L Clegg, A Mariotto, MP Fay, EJ Feuer, and BK Edwards (eds). 2003. SEER Cancer Statistics Review, 1975-2000, National Cancer Institute. Bethesda, Maryland. Accessed online in January 2004 at http://seer.cancer.gov/csr/1975_2000, 2003.

Szrom F, EG Daxon, MA Parkhurst, GA Falo, and JW Collins. 2004. *Level II and Level III Inhalation and Ingestion Dose Calculations*. PNWD-3480, prepared for the US Army by Battelle under the Chemical and Biological Defense Information Analysis Center Task 241, DO 0189, Aberdeen, Maryland.

US Army Center for Health Promotion and Preventive Medicine (USACHPPM). 2000. Depleted Uranium—Human Exposure Assessment and Health Risk Characterization in Support of the Environmental Exposure Report "Depleted Uranium in the Gulf" of the Office of the Special Assistant to the Secretary of Defense for Gulf War Illnesses, Medical Readiness and Military Deployments (OSAGWI), OSAGWI Levels I, II and III Scenarios, 15 September 2000. Health Risk Assessment Consultation No. 26-MF-7555-00D, Aberdeen Proving Ground, Maryland. Online report available at URL: www.gulflink.osd.mil in the Environmental Exposure Reports Section.

US Department of Defense (DoD). 2004. Department of Defense Deployment Biomonitoring Policy and Approved Bioassays for Depleted Uranium and Lead. HA Policy 04-004.

1.0 Introduction

The use of depleted uranium (DU) in munitions and armor has been a subject of debate and public interest since these munitions were first used by U.S. Armed Forces during the 1991 Gulf War (Operation Desert Storm [ODS]). Operation Desert Storm and subsequent military operations proved the effectiveness of armor-piercing DU cartridges and revealed the intensity of the national and international controversy generated by the use of DU. In consideration of the questions raised regarding the health risk to U.S. military personnel exposed to DU through inhalation and ingestion of aerosols generated by the perforation of Abrams tanks and Bradley Fighting Vehicles (referred to as Bradley vehicles and BFVs), the Office of the Special Assistant for Gulf War Illnesses (OSAGWI) directed the U.S. Army Center for Health Promotion and Preventive Medicine (USACHPPM) to conduct a human exposure assessment and health risk characterization for these personnel. The evaluation conducted by USACHPPM included an analysis of the available data, identification of data gaps, and determination of future research needs (USACHPPM 2000).

One of the gaps identified in the USACHPPM evaluation was the lack of robust data to support a human health risk assessment from exposure to DU aerosols for personnel in, on, or near perforated armored vehicles in which DU aerosols were generated in crew compartments. To fill this gap, OSAGWI and the U.S. Army Heavy Metals Office sponsored the Capstone DU Aerosol Study in which large-caliber (LC) DU munitions were fired at ballistic hulls and turrets (BHTs) of an Abrams tank and a Bradley vehicle. The DU aerosols generated during the Capstone study were collected and characterized. The equipment and methods used and the results of these characterizations are provided in the report entitled *Capstone Depleted Uranium Aerosol: Generation and Characterization, Volumes 1 and 2* (Parkhurst et al. 2004), referred to in this Summary Report as the Capstone DU Aerosol Study (or the Capstone study), Attachments 1 and 2. The health risks from exposures to aerosols for personnel in, on, or near perforated armored vehicles are reported in the attached report entitled *Human Health Risk Assessment of Capstone Depleted Uranium Aerosols* (Guilmette et al. 2004), referred to in this report as the Human Health Risk Assessment (HHRA), Attachment 3, and in *Level II and Level III Inhalation and Ingestion Dose Calculations* (Szrom et al. 2004), Attachment 4.

The OSAGWI (1998, 2000) and USACHPPM (2000) assessments reviewed the available relevant data contained in the public domain and in military documents. In its health risk assessment, USACHPPM staff used data from external dose measurements, DU aerosol measurements from armor impact tests, and DU oxides from burn tests of DU munitions and vehicles uploaded with DU munitions. Data from the burn tests were relatively robust in terms of the quantity of DU oxide powder produced in these fires and oxide particle characteristics. The DU oxides created from hard target impacts of penetrators into armored plate was less well characterized. In general, the experiments conducted adequately characterized the aerosols formed outside the vehicle after it was struck. However, adequate sampling of aerosols formed inside the perforated vehicle from the impact of DU munitions on non-DU and DU armor was difficult to conduct because of the violent nature of the experimental environment.

This report uses the framework developed by OSAGWI (2000) that categorizes a Soldier's potential DU exposure into three levels in which Level I has the highest potential for the highest exposures. This framework was used previously in risk assessments conducted by USACHPPM (2000) and The Royal Society (2001, 2002), and it is the basis for the HHRA (Attachment 3), which is the foundation of this Summary Report. This framework was based on interviews conducted by OSAGWI with veterans of ODS, after-action reports, and discussions with other military experts (OSAGWI 1998, 2000; USACHPPM 2000). OSAGWI staff used the interview information to develop the following three categories of exposures relative to ODS that were based on the potential for internalizing DU.

- *Level I* includes military personnel in, on, or near combat vehicles at the time of impact and perforation by DU munitions, or personnel who entered vehicles immediately after they were struck (and perforated) by DU munitions. These personnel could have been exposed to DU by fragments resulting from impact, inhalation of DU aerosols, ingestion of DU residues, or settling of DU particles on open wounds, burns, or other breaks in an individual's skin or embedded fragments—or by any combination of these possibilities. This level also includes personnel occupying a vehicle in which DU armor is perforated by non-DU munitions.

- *Level II* includes military personnel and a small number of DoD civilian employees whose job functions required them to work in and around vehicles containing DU fragments and particles. These individuals were not in the vehicle at the time of impact and did not immediately enter the vehicle after it was struck. This group performed a variety of tasks, such as battle damage assessment, repairs, explosive ordnance disposal, and intelligence gathering. They typically entered vehicles well after the initial suspended aerosol had dissipated or settled onto interior surfaces. They may have inhaled DU residues that were resuspended by their physical activities, ingested DU through hand-to-mouth transfer, or spread contamination on their clothing. DoD personnel who were involved in cleaning up DU residues generated during other events, such as the July 11, 1991, explosion and fires at the Camp Doha North Compound, are also included in this group.

- *Level III* is an "all others" group whose exposures were brief or incidental. This group includes personnel who entered DU-contaminated Iraqi equipment, were downwind from burning Iraqi or U.S. equipment struck by DU rounds, or were downwind from burning DU ammunition (e.g., personnel at Camp Doha during the July 11, 1991, explosions and fire). While these individuals could have inhaled airborne DU particles, the possibility of receiving an intake high enough to cause health effects is unlikely.

The primary differences between a Level I exposure and a Level II or Level III exposure are the delay between the armor perforation and entry of personnel into the vehicle and the length of time spent working inside the perforated vehicle. For Level I exposures during ODS, personnel within this category would have been in, on, or near the vehicle at the time of perforation, or they would have entered the vehicle soon enough after perforation to be exposed to still-suspended DU and/or DU that was resuspended from deposited material when the workers or investigators entered the vehicle. Some personnel in the Level I category could also have internalized DU through wound contamination and embedded fragments (OSAGWI 2000).

(Evaluation of the health risks from embedded fragments and wound contamination is beyond the scope of this report.)

For Level II and Level III exposures, the delay between penetration and vehicle entry of hours to days is such that exposure would be mostly from resuspended DU, hand-to-mouth transfer of the DU remaining in the vehicle, and for some categories, inhaling DU aerosols from fires involving DU munitions or armor.

1.1 PURPOSE AND SCOPE

The OSAGWI (2000) environmental exposure report and the USACHPPM report (2000) identified key data gaps for Level I exposures. This led to the initiation and completion of the Capstone study, which was designed to fill these data gaps, and the follow-on HHRA, which used the Capstone data to calculated doses and risks representative of Level I exposures. (See Attachment 1 for a complete discussion of the Capstone study.) This Summary Report is an assessment of the DU aerosol exposures to personnel during military operations. It may also be applicable to non-combatant exposures that fall within the scenario parameters. The primary purposes of this report are to:

- Update the Level I dose and risk assessments (USACHPPM 2000) based on the aerosol data collected during the Capstone study.

- Review the Level II, Level III, and Camp Doha doses and risk assessments (USACHPPM 2000) and revise, as appropriate, based on the Capstone data.

- Provide information to assist others in the revision of medical recommendations and the recommendations for protective practices as required based upon the results of the risk assessment.

1.2 APPROACH

The DU aerosol characterization and HHRA components of this overall program were contracted separately. OSAGWI and the US Army Heavy Metals Office contracted with Pacific Northwest National Laboratory to characterize the DU Aerosols. USACHPPM contracted with Battelle Memorial Institute to conduct the risk assessment. For each component, a multi-disciplinary, multi-laboratory team of health physicists, toxicologists, aerosol physicists, engineers, dose assessors, and other technical subject matter experts was assembled. The DU aerosol characterization team conducted vehicle impact tests in which armor within crew compartments was perforated to collect and characterize DU oxide compounds formed during these tests. The DU risk assessment team reviewed technical and scientific documents concerning DU, calculated DU aerosol intake and doses and concentrations to organs, and conducted a risk assessment. The risk assessment based its analysis on the results of the DU aerosol characterization documented in the Capstone study report (Attachments 1 and 2). Each major step in this process was peer-reviewed by an independent group of scientists.

This report uses the results of the HHRA (Attachment 3) as the basis for updating the Level I risk assessment, supplementing it as appropriate with information presented in USACHPPM (2000) of published DoD DU munitions test data reports. Data from the Capstone study were used to update the Level II/III intake and dose estimates where appropriate. The methodology used for estimating intakes and doses for these exposure groups was as follows:

- ***Exposure Scenarios***. Exposure scenarios were developed to model potential exposures. The actual accounts of the actions that occurred during and after friendly fire incidents during ODS were considered in the development of the exposure scenarios. These accounts were obtained by OSAGWI through interviews of crewmembers of the vehicles struck during these incidents, first responders to these incidents, and personnel who conducted battle damage assessment and recovery operations, were involved in the transport of these vehicles, or were incidentally exposed. These scenarios were used to establish critical exposure durations and the time at which exposure occurred. Adult male physiological parameters (such as breathing rate) were assigned based on the scenarios.

- ***DU Aerosol Data.*** The aerosol characteristics required to estimates intakes and doses were obtained first from the Capstone study (Attachments 1 and 2), if available, and secondly, from data presented in the DoD published test reports and summarized in USACHPPM (2000) including data from Parkhurst et al. (1995). The data generated during the Capstone study were used to estimate potential Level I exposures. Level II exposures were based on data from the Capstone study and from other reports (primarily USACHPPM 2000) generated by DoD during DU munitions testing. Evaluation of Level I exposures required a more robust data set than for Levels II or III because of their potential for higher doses. For Level I exposures, key parameters used in the dose estimates included
  —time-dependent aerosol
  —DU concentration

- particle size distribution
- solubility
- chemical composition
- particle morphology.

The results of breathing-zone air samplers worn by recovery personnel and time-dependent resuspension data from the Capstone study were used to estimate the dose for Level II personnel. Data used in the USACHPPPM (2000) report including Fliszar et al. (1989) were supplemented with Capstone data and were used to develop the dose estimates for Level III personnel. The data from many DU munitions burn tests were used to estimate the doses from fires involving DU, including data from the Camp Doha fire.

- *DU Dose Assessment.* The aerosol characteristics and the exposure scenarios provided the input parameters for the models used to estimate the dose for each of the exposure scenarios considered. Level I doses were evaluated using computer models that are based on standard, internationally accepted biokinetic and dosimetric models for both inhalation and ingestion. The data were further evaluated using Bayesian statistical analysis, which estimates parameters of an underlying (i.e., "true") distribution based on the observed distribution. The models produced radiological and chemical doses to organs of interest that allow the comparison of the doses with established safety levels and an organ-specific best estimate of the actual risk presented by the dose. A complete description of the Level I dose calculation procedures is provided in the HHRA (Attachment 3). As the data allowed, a similar procedure was implemented for Level II and Level III exposures. However, detailed modeling was not required because the estimated doses were very low.

- *Risk Assessment.* Chemical risks were evaluated through the use of a Renal Effects Risk Model, developed for this HHRA, and by comparison with the *de facto* occupational guideline for uranium kidney concentration. The Risk Model is fully described in the HHRA (Attachment 3). Radiological risks were evaluated by applying published risk coefficients to the calculated doses. The calculated risks were compared with occupational radiation risk standards. Because these standards are conservative, risks were also evaluated within the context of the hazards of combat. This second approach used a military hazard system based on the onset of effects that impair the performance of an individual or unit in combat. The HHRA (Attachment 3) has a detailed discussion of the risk assessment process used for determining Level I exposures.

1.3 DOCUMENT OVERVIEW

This Summary Report and its attachments[a] document the results of the Capstone DU Aerosol Generation and Human Health Risk Assessment Program, which focused on exposures to DU aerosols inside perforated armored vehicles. Information on exposures outside a perforated vehicle and from exposure to DU aerosol from ammunition fires is also presented within this Summary. The four reports resulting from the Capstone program are briefly described below. Additionally, the U.S. Army Medical Command (MEDCOM) developed a set of Data Quality Objectives (DQOs) for the specific information to be derived from this testing program. These DQOs are listed in Attachment 5.

Summary Report

The Summary Report includes the following elements:
- Introduction (Chapter 1)
- Characteristics of Depleted Uranium (Chapter 2), including DU properties and health risks
- Level I Exposures (Chapter 3), including modeling, source-term data, input parameters, risk assessment results, and comparison with estimates from other studies
- Level II and Level III Exposures (Chapter 4), including inhalation, ingestion, and comparison with estimates from other studies
- Depleted Uranium Chemical and Radiation Risk in the Military Context (Chapter 5)
- Putting the Risks into Perspective (Chapter 6).

DU Aerosol Generation and Characterization Report (Attachments 1 and 2)

The generation and characterization of DU aerosols in perforated Abrams tanks and Bradley vehicles is documented in the report entitled *Capstone Depleted Uranium Aerosols: Generation and Characterization, Volumes 1 and 2* (Parkhurst et al. 2004). Volume 1 of that study, hereafter referred to as the Capstone DU Aerosol Study or the Capstone study, is Attachment 1 to this Summary Report and provides the following information:
- Introduction (Chapter 1)
- Methodology and target vehicles used to create the aerosols (Chapter 2)

---
[a] Attachments 1 through 5 available on CD and Attachment 3 available in hard copy through Battelle Press. Attachments 1 and 2 also available through NTIS.

- Selection of aerosols samplers and controlling instrumentation (Chapter 3)
- Field studies in which target vehicles were fired upon and perforated using LC-DU munitions (Chapter 4)
- DU aerosol concentrations as a function of time (Chapters 5 and 6)
- Particle-size distributions as a function of time (Chapters 5 and 6)
- DU oxide concentration, composition, morphology, and dissolution in simulated lung fluid (Chapters 5 and 6)
- Conclusions (Chapter 7).

Volume 2 of the Capstone study report (Attachment 2 to this Summary Report) contains seven appendices with detailed data that resulted from the following analyses:

- DU measurement techniques and concentrations (Appendix A)
- Particle-size distributions (Appendix B)
- DU oxide composition as determined by x-ray diffraction (Appendix C)
- Particle morphology as observed using scanning electron microscopy (Appendix D)
- *In vitro* dissolution (Appendix E)
- Vehicle particle deposition surveys (Appendix F)
- Ventilation measurements of target and operational vehicles (Appendix G).

Human Health Risk Assessment (HHRA), Level I Inhalation Exposures (Attachment 3)

The evaluation of health effects of the DU aerosols measured during through the Capstone field tests is provided in *Human Health Risk Assessment of Capstone Depleted Uranium Aerosols* (Guilmette et al. 2004). That report is Attachment 3 of this Summary Report and consists of:

- Introduction (Chapter 1)
- Exposure scenarios (for Level I exposures) (Chapter 2)
- Development of input parameter values (Chapter 3)
- Chemical and radiological dose modeling (Chapter 4)
- Chemical concentrations and radiological doses (Chapter 5)
- Human health risk assessment (Chapter 6)
- Military management of DU aerosol risks (Chapter 7)

- Summary, conclusions, and recommendations (Chapter 8)
- Appendix A—Doses and DU concentration tables
- Appendix B—Results of uranium chemical toxicity in animal studies.

Level II and Level III Inhalation and Ingestion Exposures (Attachment 4)

Certain aerosol samples and deposition samples collected during the Capstone study support the analysis of exposure to residual aerosols, resuspended aerosols, and hand-to-mouth ingestion of DU. In this attachment, the process used to estimate inhalation and ingestion doses for Level II and Level III exposures is discussed and the results are presented. These results are summarized in Chapter 4 of the Summary Report.

Data Quality Objectives (DQOs) for the Capstone Study and the HHRA (Attachment 5)

The DQOs are discussed in two separate letters. The first focuses on the experimental needs to measure DU aerosol quantities and characterize the DU oxides to support the HHRA. The second focuses on the inhalation and ingestion modeling needs to estimate doses and risks for presentation in the HHRA.

1.4 REFERENCES

Fliszar, RW, EF Wilsey, and EW Bloore. 1989. *Radiological Contamination from Impacted Abrams Heavy Armor*. BRL-TR-3068, US Army Ballistic Research Laboratory, Aberdeen Proving Ground, Maryland.

Guilmette RA, MA Parkhurst, G Miller, FF Hahn, LE Roszell, EG Daxon, TT Little, JJ Whicker, YS Cheng, RJ Traub, GM Lodde, F Szrom, DE Bihl, KL Creek, and CB McKee. 2004. *Human Health Risk Assessment of Capstone Depleted Uranium Aerosols*. PNWD-3442, prepared for the US Army by Battelle under the Chemical and Biological Defense Information Analysis Center Task 241, DO 0189, Aberdeen, Maryland.

Office of the Special Assistant to the Deputy Secretary of Defense for Gulf War Illnesses (OSAGWI). 1998. *Exposure Investigation Report, Depleted Uranium in the Gulf*. Online report available at URL: www.gulflink.osd.mikl in the Environmental Exposure Reports Section.

Office of the Special Assistant for Gulf War Illnesses (OSAGWI). 2000. *Depleted Uranium in the Gulf (II), Environmental (Second Interim) Exposure Report*. Falls Church, Virginia. Online report available at URL: www.gulflink.osd.mil in the Environmental Exposure Reports Section.

Parkhurst MA, JR Johnson, J Mishima, and JL Pierce. 1995. *Evaluation of DU Aerosol Data: Its Adequacy for Inhalation Modeling*. PNL-10903, prepared for the US Army by Pacific Northwest National Laboratory, Richland, Washington.

Parkhurst MA, F Szrom, RA Guilmette, TD Holmes, YS Cheng, JL Kenoyer, JW Collins, TE Sanderson, RW Fliszar, K Gold, JC Beckman, and JA Long. 2004. *Capstone Depleted Uranium Aerosols: Generation and Characterization, Volumes 1 and 2*. PNNL-14168, Prepared for the US Army by Pacific Northwest National Laboratory, Richland, Washington.

Szrom F, EG Daxon, MA Parkhurst, GA Falo, and JW Collins. 2004. *Level II and Level III Inhalation and Ingestion Dose Calculations*. PNWD-3480, prepared for the US Army by Battelle under the Chemical and Biological Defense Information Analysis Center Task 241, DO 0189, Aberdeen, Maryland.

The Royal Society. 2001. *The Health Hazards of Depleted Uranium Munitions Part I*. Policy Document 6/01, London, United Kingdom. Online report available at www.royalsoc.ac.uk in the Science Policy Section.

The Royal Society. 2002. *The Health Hazards of Depleted Uranium Munitions Part II*. Policy Document 5/02, London, United Kingdom. Online report available at www.royalsoc.ac.uk in the Science Policy Section.

US Army Center for Health Promotion and Preventive Medicine (USACHPPM). 2000. Depleted Uranium—Human Exposure Assessment and Health Risk Characterization in Support of the Environmental Exposure Report "Depleted Uranium in the Gulf" of the Office of the Special Assistant to the Secretary of Defense for Gulf War Illnesses, Medical Readiness and Military Deployments (OSAGWI), OSAGWI Levels I, II and III Scenarios, 15 September 2000. Health Risk Assessment Consultation No. 26-MF-7555-00D, Aberdeen Proving Ground, Maryland. Online report available at URL: www.gulflink.osd.mil in the Environmental Exposure Reports Section.

2.0 Characteristics of Depleted Uranium

Over the past several decades, the U.S. military developed armor-piercing, kinetic-energy cartridges that contain depleted uranium (DU) penetrators capable of greater armor penetration than similar cartridges with tungsten penetrators. These munitions, first used by U.S. Armed Forces in combat in the 1991 Gulf War (Operation Desert Storm [ODS]), were highly effective. Their effectiveness is due to uranium's density, its tendency to "self-sharpen," and its pyrophoric nature. Depleted uranium has also been used in armor to increase the armor's effectiveness.

The radioactivity of intact DU munitions presents a minimal exposure risk to personnel. DU oxides pose a potential exposure concern if they are ingested or inhaled and subsequently deposited in the body. DU can be oxidized by burning or generation of small particles from impact with a hard target (such as an armored vehicle). As with every material, the level of risk posed is dependent upon the amount actually taken into the body.

When a DU penetrator strikes an armored target, some of its surface erodes from the rest of the penetrator. Depending on the target, the penetrator may also break into small or large fragments. Very small DU fragments and particles eroded from the surface of a DU penetrator oxidize rapidly in air and form aerosols containing DU oxides. The DU oxides and other particles from these aerosols settle out of the air onto surfaces over time. Potential health hazards may arise if a significant amount of DU oxides are inhaled or ingested. Incidental ingestion may occur from hand-to-mouth transfer.

This chapter provides a brief summary of the characteristics of uranium, DU, and DU oxides and the human health implications of DU intake.

2.1 URANIUM AND DEPLETED URANIUM PROPERTIES

Uranium is ubiquitous in the environment and is present in all soils and rocks in a variety of minerals, with granite typically having the highest concentrations. The world average uranium content of soils is about 2 μg U/g soil, and in the United States, the concentration is typically in the range of 1 to 4 μg U/g (NCRP 1976). Humans are naturally exposed to uranium primarily through drinking water. Trace amounts of uranium are found in all foods. Air concentrations of uranium also vary widely but typically are low, ranging from 0.01 to 0.2 nanogram/m^3. The uranium in air comes principally from suspended soil particles (ATSDR 1999).

Uranium is not innocuous, nor is it a "deadly poison." It is classified as a heavy metal and can cause toxicity typical of heavy metals in high doses. However, the adverse health effects are less serious than those caused by certain other heavy metals such as lead. Humans are naturally exposed to uranium from the environment without apparent health effects.

All isotopes of uranium are radioactive and chemically identical. Natural uranium consists of U-238, U-235 and U-234 with approximate mass percentages of 99.3%, 0.7%, and 0.005%, respectively. Depleted uranium is a by-product of an enrichment process that results in two product streams: one in which the U-235 and U-234 content is increased (enriched) and another in which the U-235 and U-234 content is decreased or depleted (hence the term "depleted uranium"). Depleted uranium is defined as any uranium that has less that 0.711% by mass of U-235. The U.S. military uses DU that contains approximately 0.2% by mass U-235 and is about 40% less radioactive than natural uranium. The specific activity of DoD depleted uranium is 0.38 nCi/mg (14 Bq/mg) U compared with 0.69 nCi/mg (25 Bq/mg) U for natural uranium. The DoD's DU may contain trace quantities of U-236 and contains trace quantities of transuranic elements. The dose contribution from these trace contaminants is less than 1% of the total dose (USACHPPM 2003; WHO 2001).

All isotopes of uranium emit primarily alpha particles. Radioactive decay of uranium atoms produces progeny, which also undergo radioactive decay. Beta radiation and low-level gamma radiation are emitted from the immediate short-lived progeny of U-235 (Th-231) and U-238 (Th-234, Pa-234m). Processed uranium, including DU, has had uranium chemically separated from other constituents of the ore, especially uranium progeny. DU primarily contains uranium isotopes and their immediate short-lived progeny, because sufficient time has not passed for the buildup through radioactive decay of other long-lived progeny. Table 2.1 below summarizes uranium mass percentages for each of the uranium isotopes in DoD's DU along with their decay progeny, half-lives, and radiation emissions. The types of radia-

Table 2.1. Composition of the Uranium Isotopes and Immediate Progeny in the U.S. DoD's DU

Parent	Mass %	Progeny	Half-Life	Radiation
U-238	99.8	--	4.51×10^9 yr	Alpha, Weak Gamma
		Th-234	24.1 d	Beta, Weak Gamma
		Pa-234m	1.17 min	Beta, Weak Gamma
		U-234	2.47×10^5 yr	NA[a]
U-236	0.003	--	2.4×10^7 yr	Alpha
		Th-232	1.4×10^{10} yr	NA[a]
U-235	0.2	--	7.1×10^8 yr	Alpha, Weak Gamma
		Th-231	25.6 h	Beta, Weak Gamma
		Pa-231	3.25×10^4 yr	NA[a]
U-234	0.0006	--	2.47×10^5 yr	Alpha, Weak Gamma
		Th-230	8.0×10^4 yr	NA[a]

(a) Half-lives are so long they do not contribute to the radioactivity of DU.

tion emitted from the immediate short-lived progeny are identified in the table. The immediate long-lived progeny are not significant contributors to DU's radioactivity.

The density of DU is about 1.7 times the density of lead (19.0 g/cm^3 compared to 11.4 g/cm^3). The DoD DU penetrators are 99.25% by mass DU alloyed with 0.75% by mass titanium. The improved performance of these alloyed rounds results from their higher density, their self-sharpening properties, and their pyrophoric nature. Self-sharpening means that, as the DU penetrator perforates a hard target, it forms and retains a point rather than mushrooming to a blunt shape as a tungsten penetrator does. This "adiabatic shearing" of the DU as it penetrates armor produces this self-sharpening behavior and also results in the erosion of particles that may burn (oxidize) if sufficient oxygen is present. Burning DU particles, as well as the heat generated upon perforation of a target, may ignite fuel, ammunition, or other combustible material in the target. Self-sharpening is more prevalent when the penetrator impacts the heavy armor found on tanks than when impacting lighter targets. For more information on the properties of DU, consult USACHPPM (2000) or OSAGWI (1998, 2000).

2.2 CHARACTERISTICS OF DU OXIDES GENERATED BY ARMOR PERFORATION

DU oxides can be internalized through inhalation, ingestion, or wound contamination. These oxides are formed as a result of a DU penetrator striking an armored target, any munition striking DU in armor, fires involving DU penetrators, or the corrosion of DU penetrators in the environment. Depleted uranium oxides can enter the body by inhaling aerosols generated by the initial event (perforation by a DU penetrator, perforation of DU armor, and/or a fire involving DU munitions) or by inhalation or ingestion of the oxides deposited in, on, or near the target. The physical and chemical properties of the oxide phases (U_xO_y) are important factors in dose and risk assessments because they affect particle deposition in the respiratory tract and the rate at which the uranium oxides are absorbed into the blood. USACHPPM (2000), The Royal Society (2001, 2002), and the Agency for Toxic Substances and Disease Registry (ATSDR 1999) have extensively studied and documented the general characteristics of uranium oxides.

Specific DU oxides generated by perforation of armor by a DU penetrator were characterized in the Capstone DU Aerosol Study (Attachments 1 and 2). The Capstone study field tests collected and characterized DU oxides inside target vehicles equipped with air samplers. The field tests, which incorporated various firing angles, were divided into phases based on target vehicle characteristics. These phases, in conjunction with the firing angles, allowed the investigators to separately focus on circumstances similar to ODS actions and possible future actions. In three of four test configurations (Phases I through III), the target vehicle was a ballistic hull and turret (BHT) that had been stripped of flammable material and expensive instrumentation unnecessary to the Capstone tests. These vehicles did not have any ventilation systems, and as a result, the data collected represent a reasonable upper bound of aerosols that would be expected when these vehicles were impacted. The fourth test (Phase IV) was conducted with a fully operational vehicle with a functioning ventilation system. When operating at the time of impact, the ventilation system was effective in reducing the aerosol concentrations inside the vehicle and resulted in lower estimates of intakes and doses to crewmembers.

The vehicle configurations and the number of shots fired in each of the four Capstone study field test phases are listed below and are described in Attachment 1 (Section 3.1 and Chapter 4):

- Phase I (seven shots)—Abrams tank BHT perforated through conventional (non-DU) armor, no ventilation
- Phase II (three shots)—Bradley vehicle BHT perforated through conventional armor, no ventilation
- Phase III (two shots)—Abrams tank BHT perforated through DU armor, no ventilation
- Phase IV (one DU shot; also one non-DU shot perforating armor, not addressed in this report)—Abrams tank perforated through DU armor, environmental control/nuclear, biological, chemical (EC/NBC) ventilation system operating.

Aerosol samplers selected for sample collection inside the vehicle included filter cassettes, 8-stage cascade impactors (CIs) with backup-up filters (sometimes referred to as 9 stages), a 5-stage cyclone separator, and a moving filter (Attachment 1, Section 3.5). Limited sampling outside the vehicle was also conducted using high-volume air samplers in the tests (Phase I, Shot 1 through Phase I, Shot 4) and cascade impactors for the remaining tests (Attachment 1, Section 3.6). Gloves and wipe-test survey samples provided additional data for use in evaluating incidental DU ingestion from hand-to-mouth transfer (Attachment 1, Sections 3.7). The concentrations of DU in the aerosols as a function of time for each shot and each sampling position are provided in Attachment 1, Sections 5.1 and 6.1.

Generalizations related to the physical and chemical characteristics of the aerosols as collected inside the target vehicles by the 5-stage cyclone or the 9-stage cascade impactors include:

- The percentages of uranium mass in the total mass of aerosol collected in the cyclone samples varied (Attachment 1, Sections 5.6 and 6.4):
  —38 to 72% in the Abrams tank perforated through conventional armor
  —60 to 72% in the Abrams tank perforated through DU armor
  —18 to 29% in the Bradley vehicle perforated through conventional armor.

  The uranium percentages decreased slightly with decreasing particle size in the data from the Abrams tank (with and without DU armor) samples and increased slightly with decreasing particle size in the Bradley vehicle samples.

  Most of the remaining aerosol mass consisted of oxygen, aluminum, and iron. Very small or trace amounts of other metals, including titanium (which was alloyed with DU in the penetrator), zinc, and copper were also present in the aerosol. The aluminum content varied the most by test phase; it was highest in the Phase II samples and lowest in the Phase III samples.

- Total activity and uranium mass concentrations decreased rapidly as a function of time as settling occurred. Within 30 min, concentrations were typically about two orders of magnitude lower than concentrations within the first few seconds (Attachment 1, Chapters 5 and 6).

- Particle size distributions changed as a function of time as settling occurred. The particle size distributions were analyzed using activity median aerodynamic diameters (AMADs) and fitting unimodal and bimodal distributions. The bimodal model provided a better fit for many of the samples, but neither approach adequately described a significant number of samples (Attachment 1, Sections 5.5 and 6.2.2;

Attachment 2, Appendix B). Consequently, for the Human Health Risk Assessment (HHRA), the DU aerosol distribution was represented by nine monodisperse aerosols (Guilmette et al. 2004, Attachment 3). The nine monodisperse aerosol concentrations were derived from the amounts of DU deposited on the nine stages of the CIs. Using this approach to represent the particle-size distribution more accurately describes the actual size distribution data and is preferable to using fitted AMADs (discussed in Attachment 3, Section 4.4.1).

- The predominant uranium oxide phase consisted of U_3O_8/UO_3. Its fraction increased as particle size decreased while the fraction of U_4O_9, which was highest in the large particles, decreased as particle size decreased. A small amount of $UO_3 \cdot 2H_2O$ (schoepite) was detected in several cyclone stages and in backup filter samples (Attachment 1, Sections 5.6.5 and 6.43; Attachment 2, Appendix C).

- The particles obtained from the cyclone samples and from other samples evaluated under a scanning electron microscope showed that the aerosols had a complex, heterogeneous structure. The uranium particles displayed many different shapes, ranging from spherical shapes to granular-appearing structures to fractured appearances. These different shapes suggest the likelihood that the DU particles were formed by several different mechanisms (Attachment 1, Sections 5.6.6 and 6.4.4; Attachment 2, Appendix D).

- The in vitro dissolution rates of the cyclone residues tended to increase with decreasing particle size, but the relationship was not strong. It is reasonable to assume that the observed variability was attributable, at least in part, to the significant particle heterogeneity. The samples had dissolution rates that most closely resembled Type M (a moderate rate of dissolution) absorption behavior with the exception of one backup filter, which resembled Type S (slowly dissolving) behavior (based on the Publication 66 Human Respiratory Tract Model, International Commission on Radiological Protection (ICRP [66] 1994]). More than half of each sample (about 58 to 99%) fit the Class-Y (half-time greater than 100 days) clearance category described in Publication 30 (ICRP [30] 1979) (Attachment 1, Sections 5.6.7 and 6.4.5; Attachment 2, Appendix E). Several samples also contained a Class W (intermediate) component. Actual measured dissolution rates rather than dissolution types were used in dose modeling.

Resuspended aerosols were collected during sample recovery activities. Potential hand-to-mouth transfer of deposited DU material was evaluated using measurements of DU contamination on cotton gloves worn by members of the sample recovery team. These data were supplemented by interior vehicle wipe survey results and subsequently used in dose modeling for potential Level I and Level II exposures (Attachment 1, Sections 5.4 and 6.3.3; Attachment 2, Appendix F; and Attachment 4).

The external air samplers yielded uranium concentrations and particle-size distributions (Attachment 1, Section 5.3). Because the aerosols generated by target perforation were confined within the Aberdeen Test Center Superbox, use of these data to estimate exterior aerosol source terms will overestimate DU aerosol concentrations as compared to battlefield conditions.

2.3 HEALTH RISKS OF DEPLETED URANIUM

DU oxides from armor perforation, fires involving penetrators, or natural erosion of DU penetrators in the environment are internalized through inhalation, ingestion, and wound contamination. DU oxides can be inhaled and internalized at the time of the initial event (perforation of armor by a DU penetrator, perforation of DU armor, and/or the fire involving DU munitions) as a result of aerosols generated. They can also be internalized after the event as a result of inhalation (resuspension) or ingestion of the oxides deposited on the vehicle or in the environment (primarily soil) surrounding the vehicle after the event. The interaction of the DU penetrator with armor creates DU oxides and aggregates with iron, aluminum, or other material components of the armor (most of which would be trace amounts). Projecting potential health effects for personnel who inhale these DU oxides from armor perforations is the focus of this HHRA. The potential contributing effect of inhaling the non-DU particles is outside the scope of this evaluation.

The chemical toxicity of uranium compounds is well known. In 1824, a treatise described uranium salts as "feeble poisons" when given by mouth to animals (cited in Hodge et al. 1973). In the late 1800s, uranium salts were studied as homeopathic therapeutic agents in humans, primarily for diabetes. In the early 1900s, the renal toxicity of uranium became apparent in humans, and its use as a therapeutic agent ceased. Intense radiological and chemical toxicity studies were initiated during World War II as nuclear weapons were developed. After the war, the U.S. Atomic Energy Commission supported continuing studies of the toxicity of uranium and other compounds related to nuclear weapons.

A comprehensive review of the hazards of uranium and protection criteria was published in 1973 (Hodge et al. 1973). More recently, reviews of the toxicity (ATSDR 1999) and carcinogenicity (BEIR IV 1988) of uranium have been published, as well as several reviews of the toxicity of DU used in munitions (Harley et al. 1999; Fulco et al. 2000; WHO 2001; The Royal Society 2001, 2002). The HHRA contains an abbreviated summary of the pertinent data (Attachment 3, Chapter 6 and Appendix B).

2.3.1 Chemical Health Risks

Uranium has been demonstrated to affect several organ systems following exposure to sufficiently high concentrations; the most important of these organs are the kidneys and the lungs (ATSDR 1999). As with any chemical, the dose, or concentration in the tissues, is the primary factor in the chemical toxicity of uranium. This dose is determined by the amount internalized and route of internalization as well as the solubility of the chemical form of uranium compound. Inhalation is the most common route of occupational exposure to uranium. Solubility dictates how rapidly the compound moves from the lungs to the blood and then to other organs. Uranium oxides are the principal forms of uranium found following impacts of DU munitions with armor, and using default International Commission on Radiological Protection (ICRP) solubility classifications would be classified as Type M or Type S, indicating moderate and slow dissolution from the lung (ICRP [66] 1994). In this HHRA, inhalation of the relatively insoluble uranium oxides generated by impacts of DU munitions with armor is assumed to be the primary route of exposure for vehicle occupants. Incidental hand-to-mouth ingestion is also possible and is discussed in Chapter 4. Actual measured dissolution rates from the Capstone study rather than the ICRP default dissolution types were used in dose modeling. It should be noted that vehicle occupants could also be exposed through wound contamination or embedded fragments. Estimating the dose from these routes is beyond the scope of this work.

Once uranium has solubilized and entered the bloodstream, the kidney is considered to be the target organ for chemical effects. The value of 3 µg/g kidney tissue is the *de facto* occupational guideline for the maximum permissible concentration of uranium in the kidney, and is based on the results of occupational exposures to uranium, the results of animal inhalation experiments (Spoor and Hursch 1973), and extrapolation from radiation standards (ICRP [2] 1959). The HHRA has a detailed review of the chemical health risks associated with internal DU exposure (Attachment 3).

In addition to reviewing animal data during the course of this assessment, data from human studies were reviewed. These data included 14 cases previously reviewed by The Royal Society (2002) of renal effects from acute uranium exposures and 13 additional cases described by Fisher et al. (1990). Data from these studies were used to develop a model correlating uranium kidney concentrations to categories of health effects. Rather than providing a single guideline such as the *de facto* occupational guideline, this model divides kidney concentrations into four categories termed Renal Effects Groups (REG), based on the severity of the renal effects and the predicted outcome (Table 2.2). Details of how this model was developed are contained in Attachment 3, Chapter 6 and Appendix B. Chemical effects to other organs are also discussed in Attachment 3, Chapter 6.

Table 2.2. Renal Effects Groups—Chemical Risk to the Kidneys

Renal Effects Group	Uranium Kidney Concentration (µg U/g kidney tissue)	Acute Renal Effect	Predicted Outcome
0	≤2.2	No Detectable Effects	No Detectable Effects
1	>2.2 to ≤6.4	Possible Transient Indicators of Renal Dysfunction	Not Likely to Become Ill
2	>6.4 to ≤18	Possible Protracted Indicators of Renal Dysfunction	May Become Ill
3	>18	Possible Severe Clinical Symptoms of Renal Dysfunction	Likely to Become Ill

2.3.2 Radiological Health Risks

The primary radiological health risk for inhaled depleted uranium is a potential for an increase in cancer. There is, however, no compelling evidence from human epidemiologic studies that uranium is associated with an increased cancer risk. The concern that DU may increase cancer is based on the knowledge that radiation doses can be delivered to various organs of the body from inhaled DU and that radiation is a known carcinogen. Biokinetic model calculations indicate that the risk of internally deposited DU is primarily to the lung. Increased mortality based on risk coefficients of alpha-emitting radionuclides was used to calculate risk of cancer to selected organs. Radiation doses to organs and their corresponding estimates of increased cancer mortality from inhaled DU are provided in the HHRA (Attachment 3, Sections 6.5 and 6.6). Calculated cancer mortality risks at the 10th and 90th percentile are reported for the lung and kidney in Chapter 3.

2.3.2.1 Effects in the Lungs

Irradiation of the lungs has been convincingly associated with lung cancer, but not from uranium exposure. The National Academy of Sciences (NAS), Institute of Medicine, recently reviewed 11 epidemiologic studies of workers in six major uranium-processing facilities in the United States. The NAS committee concluded from these 11 epidemiologic studies that there was "limited evidence indicating no association between exposure to uranium and lung cancer at cumulative doses lower that 200 mSv [20 rem]" and that there was "insufficient evidence to determine an association at higher doses" (Fulco et al. 2000). For purposes of the HHRA, the ICRP 60 (1991)/NCRP (1993) lung cancer risk coefficient for worker populations was used in this modeling effort.

2.3.2.2 Other Organs

There is no direct evidence that natural uranium will cause bone cancers in humans, although enriched uranium will induce bone sarcomas in rats and mice (ATSDR 1999). In its review of epidemiology studies of workers in uranium processing facilities in the United States, the NAS, Institute of Medicine, concluded that there was insufficient evidence to determine an association between uranium exposure and bone cancer (Fulco et al. 2000).

Bone marrow might be considered a target organ because of the known deposition of uranium in the bone. However, the relevant epidemiologic experiences with alpha-emitting radionuclides in human bone have been reviewed with the conclusion that the risk from radiation-induced leukemia is insignificant relative to bone sarcoma (Mays et al. 1985).

The thoracic lymph nodes potentially accumulate large burdens of insoluble uranium compounds if lung burdens are large. However, there are no data from studies of humans to suggest that large burdens might result in cancers of the lymph nodes. In fact, studies of exposures to dogs suggest that cancers do not develop in the lymph nodes after high burdens and radiation dose from uranium dioxide inhaled for 5 years (Leach et al. 1973). In addition, the NAS, Institute of Medicine, in its review of epidemiology studies of workers in uranium processing facilities in the United States concluded that there was insufficient evidence to determine an association between uranium exposure and lymphatic cancer (Fulco et al. 2000).

Further discussion of these and other organs is provided in Attachment 3, Section 6.6. With the exception of the kidney, ICRP 60 (1991) risk coefficients for fatal cancers were used in the analysis. The coefficient for kidney risk is from the U.S. Environmental Protection Agency's Geometric Mean Coefficient Model (Puskin and Nelson 1995).

2.3.3 External Exposures

Depleted uranium decays by emission primarily of alpha and beta particles and a very small fraction of gamma rays. Intact DU munitions and undamaged DU armor represent no risk of contamination and emit low levels of penetrating radiation. When handling bare penetrators, gloves should be worn to shield the non-penetrating radiation. There have been extensive and definitive studies of the external exposures from DU in a variety of configurations, ranging from single cartridges and packaged rounds to ammunition pallets and bulk storage facilities to rounds loaded in combat vehicles (both in Abrams tanks [with and without DU armor] and in Bradley vehicles). These results are discussed in USACHPPM (2000) and OSAGWI (2000).

2.4 REFERENCES

Agency for Toxic Substances and Disease Registry (ATSDR). 1999. *Toxicological Profile for Uranium.* Report TP 90-29, Atlanta, Georgia.

BEIR (Biological Effects of Ionizing Radiations) IV. 1988. *Chapter 6. Uranium* In *Health Risks of Radon and Other Internally Deposited Alpha-Emitters.* Committee on the Biological Effects of Ionizing Radiations, National Research Council, National Academy Press, Washington DC.

Fisher DR, MJ Swint, and RL Kathren. 1990. "Evaluation of Health Effects in Sequoyah Fuels Corporation Workers from Accidental Exposure to Uranium Hexafluoride." NUREG/CR-5566, US Nuclear Regulatory Commission, Washington, DC.

Fulco CE, CT Liverman, and HC Sox, eds. 2000. "Depleted Uranium." In *Gulf War and Health*, Vol. 1, Chapter 4, National Academy Press, Washington, DC.

Guilmette RA, MA Parkhurst, G Miller, FF Hahn, LE Roszell, EG Daxon, TT Little, JJ Whicker, YS Cheng, RJ Traub, GM Lodde, F Szrom, DE Bihl, KL Creek, and CB McKee. 2004. *Human Health Risk Assessment of Capstone Depleted Uranium Aerosols.* PNWD-3442, prepared for the US Army by Battelle under Chemical and Biological Defense Information Analysis Center Task 241, DO 0189, Aberdeen, Maryland.

Harley NH, FC Foulkes, LH Hilborne, A Hudson, and CR Anthony. 1999. *A Review of the Scientific Literature As It Pertains to Gulf War Illnesses: Volume 7, Depleted Uranium.* MR-1018/7-OSD, RAND, Santa Monica, California. Online report available at URL: www.gulflink.osd.mil in the Environmental Exposure Reports Section.

Hodge HC, JN Stannard, and JB Hursh, eds. 1973. "Uranium, Plutonium, Transplutonium Elements." *Handbook of Experimental Pharmacology*, Vol 36, Springer-Verlag, New York.

International Commission on Radiological Protection (ICRP). 1959. *Report on the Committee II on Permissible Dose for Internal Radiation.* ICRP Publication 2, Pergamon Press, Oxford, United Kingdom.

International Commission on Radiological Protectection (ICRP). 1979. *Limits for Intakes of Radionuclides by Workers.* ICRP Publication 30, Part 1, Pergamon Press, Oxford, United Kingdom.

International Commission on Radiological Protection (ICRP). 1991. *Recommendations of the International Commission on Radiological Protection.* ICRP Publication 60, Pergamon Press, Oxford, United Kingdom.

International Commission on Radiological Protection (ICRP). 1994. *Human Respiratory Tract Model for Radiological Protection.* ICRP Publication 66, Pergamon Press, Oxford, United Kingdom.

International Commission on Radiological Protection (ICRP). 1995. *Age-dependent Doses to Members of the Public from Intake of Radionuclides: Part 4 Inhalation Dose Coefficients.* ICRP Publication 71, Pergamon Press, Oxford, United Kingdom.

Leach LJ, CL Yuile, HC Hodge, GE Sylvester, and HB Wilson. 1973. "A Five-Year Inhalation Study with Natural Uranium Dioxide (UO_2) Dust-II. Postexposure Retention and Biologic Effect in the Monkey, Dog, and Rat." *Health Physics* 25:239-258.

Mays CW, RE Rowland, and AF Stehney. 1985. "Cancer Risk from the Lifetime Intake of Ra and U Isotopes." *Health Physics* 48:635-647.

National Council on Radiation Protection and Measurements (NCRP). 1976. *Environmental Radiation Measurements.* NCRP Report No. 50, Washington, D.C.

National Council on Radiation Protection and Measurements (NRCP). 1993. *Risk Estimates for Radiation Protection.* NCRP Report No. 115, Bethesda, Maryland.

Office of the Special Assistant to the Deputy Secretary of Defense for Gulf War Illnesses (OSAGWI). 1998. *Exposure Investigation Report, Depleted Uranium in the Gulf.* Online report available at URL: www.gulflink.osd.mil in the Environmental Exposure Reports Section.

Office of the Special Assistant to the Deputy Secretary of Defense for Gulf War Illnesses (OSAGWI). 2000. *Exposure Investigation Report, Depleted Uranium in the Gulf (II).* Online report available at URL: www.gulflink.osd.mil in the Environmental Exposure Reports Section.

Puskin JS and CB Nelson. 1995. "Estimates of Radiogenic Cancer Risks." *Health Physics* 69:93-101.

Spoor NL, and JB Hursch. 1973. "Protection Criteria." In "Uranium, Plutonium, Transplutonium Elements." Hodge HC, JN Stannard, JB Hursh, eds., pp. 241-269, *Handbook of Experimental Pharmacology,* Vol 36, Springer-Verlag, New York.

The Royal Society. 2001. *The Health Hazards of Depleted Uranium Munitions Part I.* Policy Document 6/01, London, United Kingdom. Online report available at URL: www.royalsoc.ac.uk in the Science Policy Section.

The Royal Society. 2002. *The Health Hazards of Depleted Uranium Munitions Part II,* Policy Document 5/02, London, United Kingdom. Online report available at URL: www.royalsoc.ac.uk in the Science Policy Section.

US Army Center for Health Promotion and Preventive Medicine (USACHPPM). 2000. *Depleted Uranium—Human Exposure Assessment and Health Risk Characterization in Support of the Environmental Exposure Report "Depleted Uranium in the Gulf" of the Office of the Special Assistant to the Secretary of Defense for Gulf War Illnesses, Medical Readiness and Military Deployments (OSAGWI), OSAGWI Levels I, II and III Scenarios, 15 September 2000.* Health Risk Assessment Consultation No. 26-MF-7555-00D, Aberdeen Proving Ground, Maryland. Online report available at URL: www.gulflink.osd.mil in the Environmental Exposure Reports Section.

US Army Center for Health Promotion and Preventive Medicine (USACHPPM). 2003. Memorandum for Record, MCH-TS-OHP, 14 February 2003, Subject: *Transuraniucs (TRUs) and Technicium-99 (Tc 99) Contaminants in Depleted Uranium Armor.*

World Health Organization (WHO). 2001. *Depleted Uranium: Sources, Exposure, and Health Effects.* Department of the Protection of the Human Environment, Geneva, Switzerland. Online report available at URL: www.who.int/environmental_information/radiation/depleted_uranium.htm.

3.0 Level I Exposures

Level I exposures encompass personnel in, on, or near combat vehicles at the time these vehicles were struck (and perforated) by depleted uranium (DU) munitions, or personnel who entered vehicles immediately after impact and perforation by DU munitions. Intake of DU by these personnel could have resulted from DU fragment wounds, inhaled DU aerosols, ingested DU residues, or DU particles deposited on open wounds, burns, or other breaks in the skin—or by any combination of these possibilities.

The Human Health Risk Assessment (HHRA) (Guilmette et al. 2004, Attachment 3) provides the most definitive assessment to date of Level I exposures using the data provided by the Capstone DU Aerosol Study (Parkhurst et al. 2004, Attachments 1 and 2). This chapter provides a summary of inhalation exposure scenarios, risk assessment parameters and assumptions, and the HHRA results, which are arranged by vehicle configuration, exposure start time, and duration of exposure. It also compares the HHRA results with risk assessments conducted for Level I exposures by USACHPPM (2000) and The Royal Society (2001, 2002). Exposure to DU ingested via hand-to-mouth transfer would be similar to Level II ingestion exposures discussed in Chapter 4.

3.1 ODS CREWMEMBER EXPOSURE INCIDENTS

According to DoD, Force Health Protection, formerly known as the Office of the Assistant Secretary of Defense of the Gulf War Illnesses (OSAGWI), six Abrams tanks were involved in friendly fire incidents during Operation Desert Storm (ODS), and in three of those six cases, the crew compartments were perforated. Most of the surviving crew exited quickly or were outside at the time of the incident. The ventilation system was operating in at least one of the three cases, and the Halon system, which would tend to reduce the DU aerosol, activated in these three tanks. Open hatches also would have helped disperse the aerosol and reduced doses. This information suggests that a maximum of 10 Soldiers (all male) were exposed to DU aerosol conditions similar to, but less severe than those simulated during the Capstone Phase I field tests, in which a large-caliber DU munition perforated an Abrams tank through conventional armor.

Similarly, fifteen Bradley Fighting Vehicles (Bradley vehicles) were perforated through crew compartments by large-caliber DU munitions. Bradley crewmembers (all male) were inside the perforated vehicles in 14 of the 15 perforations. Fewer details are available about the exiting times and activation of ventilation in these incidents. Some of these vehicles were hit more than once, and the Halon fire suppression system was activated in at least a few cases. If fire-suppression systems in a vehicle are activated, it is crucial that crewmembers quickly exit to prevent suffocation.

Abrams surviving crewmembers are believed to have exited within two minutes after perforation. In most cases, the exit times for Bradley vehicle crewmembers are assumed to be similar to the exit times for Abrams tank crewmembers. However, some of the Bradley vehicles could be driven after perforation, and these vehicles were driven to repair stations. Therefore, longer stay-times may be applicable for some personnel in the Bradley vehicles.

It was a common practice during ODS to open hatches on perforated vehicles to reach wounded personnel inside the vehicle so they could be treated and evacuated. The scenarios assumed that crewmembers would take no action to clear the air in the vehicle, such as activating the ventilation system or opening hatches for the duration of the stay-time. While this may be reasonable for the 1- and 5-min stay times, it is less reasonable for the longer stay-times.

3.2 MODELING PARAMETERS AND ASSUMPTIONS

Exposure start times and durations, physiological parameters, and the source term with respect to time were modeled for the DU aerosol concentrations, intakes, and doses. The exposure start times and durations were used in developing scenarios for which results are presented. A series of reports by the International Commission on Radiological Protection (ICRP) provided physiological parameters based on "Reference Man" characteristics (ICRP

[23] 1975, [66] 1994a, [70] 1995a, [89] 2003). Source term data were taken from the Capstone study (Attachments 1 and 2).

3.2.1 Inhalation Exposure Scenarios

Five scenarios were developed to represent Level I exposures for DU aerosols inhaled within the perforated vehicles (Attachment 3, Chapter 2). The first four, Scenarios A through D, were used to model exposures to personnel in the vehicle at the time it was perforated. These scenarios vary by exposure duration (length of time before vehicle exit) and the breathing rate of the personnel involved. The fifth scenario (E) models exposure to first responders. For this scenario, these personnel were not in the vehicle at the time of perforation but entered shortly after perforation. For exposure durations up to 15 min, a breathing rate consistent with a person engaged in heavy exercise (3 m^3/h) was used in the model. For longer exposure durations, a breathing rate consistent with a person engaged in light exercise (1.5 m^3/h) was used. The scenarios assume no respirator use.

Scenarios for Individuals Inside the Vehicles during Vehicle Perforation

Four exposure durations were modeled for personnel inside the vehicles at the time of perforation: 1 min (Scenario A), 5 min (Scenario B), 1 h (Scenario C), and 2 h (Scenario D). The first two exposure durations assumed that the crew would be able to readily exit the vehicle. Scenarios A and B closely resemble the ODS experiences in which surviving crewmembers reported exiting Abrams tanks quickly, (Deployment Health Support Directorate 2004) in which all who could exit were believed to have exited in less than 5 min. These are identified as the most likely scenarios. Scenarios C and D were used to model circumstances for which the vehicle continued to operate post-perforation, and therefore the crew may have had longer exposure durations. These are identified as the upper bound scenarios. During this time period, the overall DU aerosol concentration decreases. Longer stay-times would increase exposure, but the DU aerosol concentrations after 2 h are orders of magnitude lower than those occurring immediately after perforation and therefore add incrementally less to DU intake.

Scenario for First Responders

It was assumed that first responders (Scenario E) would enter the struck vehicle soon after vehicle perforation, usually to recover personnel or equipment. These individuals would NOT be in or on the combat vehicle during perforation. Their exposures would depend primarily on how soon after perforation they enter the vehicle and secondarily on how long they remain inside. For purposes of modeling, these individuals were assumed to enter the vehicle 5 min post-perforation to help evacuate the vehicle occupants and then exit the vehicle 10 min later. Their breathing would be at rates consistent with heavy exercise (3 m^3/h) during that time period. It was assumed that these entry personnel would not wear respiratory protection. A summary of the key exposure scenario parameter values is provided in Table 3.1.

Table 3.1. Summary of Level I, In-Vehicle Personnel Exposure Scenarios

Scenario	Time of Exposure	Exposure Duration	Breathing Rate
Crew Inside Vehicle			
A	From impact to exit 1 min post shot	1 min	3 m^3/h
B	From impact to exit 5 min post shot	5 min	3 m^3/h
C	From impact to exit 1 h post shot	1 h	3 m^3/h for first 15 min, 1.5 m^3/h thereafter
D	From impact to exit 2 h post shot	2 h	3 m^3/h for first 15 min, 1.5 m^3/h thereafter
First Responder			
E	Entry 5 min post shot, exit 10 min later	10 min	3 m^3/h

3.2.2 Physiological Parameters

The Level I exposure cohort from ODS consisted only of adult males; consequently, modeling was limited to male physiological parameters that have been standardized into the "Reference Man" characteristics by the International Commission on Radiological Protection (ICRP). The characteristics relevant to the respiratory tract are based on ICRP 66 (1994a), those related to bone are based on ICRP 70 (1995a), and those related to the remainder of the organs are from ICRP 23 (1975) and ICRP 89 (2003).

Breathing rate and type (route) of breathing also affect dose. Breathing rates are related to type and amount of physical exertion. ICRP 66 default values for a healthy adult male (i.e., Reference Man) were used assuming light and heavy exercise conditions. The path of inhalation entry, whether through the nose or mouth, affects the aerosol deposition patterns in the respiratory tract. Most people normally breathe through their noses, particularly at lower physical activity levels, and the default case used in these calculations was for a nose-breathing individual.

A discussion of these and other physiological parameters is provided in Attachment 3, Chapter 3.

3.3 CHEMICAL AND RADIOLOGICAL DOSE MODELS

Determining the concentration of uranium and the radiation dose in tissues and organs as a function of time requires the use of models that describe the following processes:

- intake and deposition of the uranium into the respiratory tract
- clearance from the respiratory tract to the gastrointestinal tract
- removal from the gastrointestinal tract either to the blood or by elimination from the body
- removal from the respiratory tract to the blood and lymphatic system
- distribution through blood and lymphatic system to other tissues and organs of the body concurrent with elimination through the urinary tract
- retention in the blood-fed tissues and organs and feedback from those tissues and organs to blood.

The ICRP has developed several models to mathematically describe these processes. The following three models, which are the most recent models published by ICRP, were selected for the radiological dose assessment documented in the HHRA:

- the inhalation/respiratory tract model described in ICRP 66 (1994a)
- the ingestion/gastrointestinal tract model described in ICRP 30 (1979)
- the uranium systemic biokinetic model in ICRP 69 (1995b).

These models and their implementation are discussed in Attachment 3, Chapter 4. The data derived from applying these models include intake amounts, organ-specific DU mass concentrations and doses as a function of time, and time-integrated doses for each of the scenarios.

3.4 AEROSOL CHARACTERISTICS AND SOURCE-TERM DATA

Thirteen field tests were conducted during the Capstone study, and aerosols were collected and characterized for aerosol composition, particle size, morphology, and *in vitro* dissolution. Results from these tests provide a basis for bracketing the range of expected aerosol parameter values. Time-dependent aerosol concentrations and particle-size distributions as measured on 8-stage cascade impactors (CIs) were used as the primary source-term data for modeling. Time-integrated samples collected in a five-stage cyclone sampler were evaluated for aerosol composition, morphology, and *in vitro* dissolution. Samplers used in collecting these aerosol samples and the methods used to characterize the aerosols are discussed in Attachment 1, Section 3.5, and the results of these analyses are presented in Attachment 1, Chapter 5, and in Attachment 2.

In a departure from the traditional practice of using either assumed or fitted unimodal or bimodal lognormal particle-size distributions, actual uranium masses from each of the eight size-specific CI stages and backup filter were used directly in the dose assessments. This approach better maintained the quality and variability of the original measurement data and avoided the introduction of additional uncertainty into the dose assessment from poor data fits. Likewise, *in vitro* solubility data from time-integrated cyclone samplers were used to define the rates of DU absorption to blood in a size- and material-specific manner. The computational approach used to implement these data into the modeling is discussed in Attachment 3, Section 4.4.

Using the results of the uranium collected on each CI stage, median intakes of DU were predicted from the aerosols for the scenarios. These intakes are listed for each vehicle configuration in Table 3.2. The intakes in the most likely scenario range from a low of 10 mg for a 1 min exposure in the ventilated Abrams tank with DU armor to a high of 710 mg for a 5 min exposure in an unventilated Abrams tank with DU armor. These data were used as input to the models calculating DU concentrations to tissues and organs and radiological doses.

Table 3.2. Summary of HHRA Median DU Intakes

Scenarios	Uranium Intake (mg)			
	Abrams Tank: Conventional Armor, No Ventilation	Abrams Tank: DU Armor, No Ventilation	Abrams Tank: DU Armor, EC/NBC Operating	Bradley Vehicle: Conventional Armor, No Ventilation
Most Likely				
A - Crew, exit in 1 min	280	250	10	83
B - Crew, exit in 5 min	590	710	43	220
E - First responders	160	200	27	99
Upper Bound				
C - Crew, exit in 1 h	760	970	91	330
D - Crew, exit in 2 h	780	1000	110	380

3.5 CAPSTONE DOSE AND RISK ASSESSMENT RESULTS

The HHRA (Attachment 3) used the Capstone study aerosol results (Attachments 1 and 2) to develop estimates of DU intakes, tissue/organ DU concentrations, and radiological doses. These estimates were further used in evaluating chemical toxicity to the kidney and radiological risk. Statistical analyses of the Phase I through III data sets were used to calculate the mean, median, and 10th and 90th percentiles (presented in Attachment 3, Chapter 6). (Phase IV data were based on four single time-sequenced samples.) The data in Attachments 1 and 3 are typically presented to three significant figures. In this Summary Report, the results are presented to two significant figures or to the number of decimals most useful for data presentation.

The kidney is the principal target organ for uranium toxicity. Therefore, the HHRA based its estimate of risk on a comparison of the projected uranium concentrations in the kidney with the human data available regarding toxicity of uranium to the kidney. These human data were used to develop a Renal Effects Grouping (REG) system that is based on clinical manifestations of kidney toxicity. This approach simplifies the interpretation of the kidney data that were obtained.

The HHRA acknowledges that the long-term radiation health effects of inhaled uranium compounds have been difficult to identify with any certainty in human populations. In fact, natural uranium, which is more radioactive than DU, has not been shown to cause cancer in occupationally exposed workers, although epidemiologic studies of workers in the uranium processing industry suggest that there is a very weak association between exposure to inhaled uranium and cancer (Fulco et al. 2000). Modeling results provided committed effective doses, E(50); tissue/organ committed equivalent doses, $H_T(50)$; and selected DU tissue/organ concentration versus time profiles. Estimates of risk were based on the summation of the individual risks from each tissue/organ (organ dose times the organ risk) rather than estimating risk based upon a calculation of the whole body effective dose. This was done because of the non-uniformity of the dose from inhaled DU with 80 to 90% of the risk attributable to the lung exposure and the availability of risk coefficients for alpha radiation of the lung. Tissue- and organ-specific risks for the other organs were either obtained from the literature or inferred from the tissue-weighting factors in ICRP 60 (1991) and ICRP 68 (1994b). (The dose to the extrathoracic region was high enough to warrant assigning a weighting factor of 0.025.) In all cases, the summed tissue/organ risks were a factor of approximately 1.35 higher than the risk estimated using the whole body effective dose.

3.5.1 Capstone HHRA Chemical Toxicity Results

Table 3.3 summarizes the median peak kidney uranium concentrations for each scenario in each vehicle configuration tested. The most likely scenarios yielded median peak kidney uranium concentrations for crewmembers in an Abrams tank that ranged from 0.05 µg U/g kidney (Scenario A for an Abrams

Table 3.3. Summary of HHRA Median Peak Kidney Uranium Concentrations

Scenarios	Peak Kidney Uranium Concentrations (µg U/g Kidney)			
	Abrams Tank: Conventional Armor, No Ventilation	Abrams Tank: DU Armor, No Ventilation	Abrams Tank: DU Armor, EC/NBC Operating	Bradley Vehicle: Conventional Armor, No Ventilation
Most Likely				
A - Crew, exit in 1 min	3.0	1.1	0.05	1.0
B - Crew, exit in 5 min	6.4	2.6	0.23	2.9
E - First responders	1.5	0.67	0.14	1.4
Upper Bound				
C - Crew, exit in 1 h	8.2	3.5	0.46	3.5
D - Crew, exit in 2 h	8.0[a]	3.7	0.56	4.0

(a) The sampler data used to calculate Scenarios C and D differed slightly and was responsible for the lower dose for the longer exposure time.

tank with DU armor perforated by a DU munition and with the environmental control/nuclear, biological, chemical (EC/NBC) ventilation system operating) to 6.4 µg U/g kidney (Scenario B for perforation of an Abrams tank without DU armor and with the EC/NBC ventilation system not operating). Bradley vehicle scenarios yielded results similar to the unventilated Abrams tank with DU armor. For reference, the *de facto* occupational guideline is 3 µg U/g kidney.

The levels in the most likely scenarios represent a range of REG 0 (below 2.2 µg U/g kidney) to the interface between REG 1 and 2 (6.4 µg U/g kidney; REG ranges listed in Table 2.2). Kidney concentrations within REG 0 and 1 are unlikely to cause illness. Concentrations within REG 2 may cause temporary kidney effects. Median kidney uranium concentrations predicted for upper bound scenarios ranged from 0.46 µg U/g kidney (REG 0) to 8.2 µg U/g kidney (REG 2).

The results predict that kidney uranium concentrations would be greater from perforating conventional Abrams armor perforation than from perforating DU armor of an Abrams tank. This is due mainly to the slower dissolution rates for the DU oxides generated in the DU armor perforation tests.

The impact of the operating EC/NBC ventilation system on aerosol removal is demonstrated by the reduced kidney uranium concentrations of at least a factor of four. Although not field tested, calculations in the HHRA estimate that a significant concentration reduction would also occur with operation of ventilation systems on the Bradley vehicle.

Predicted median peak kidney uranium concentrations are presented by vehicle configuration and scenario in Figure 3.1. The medians values are represented

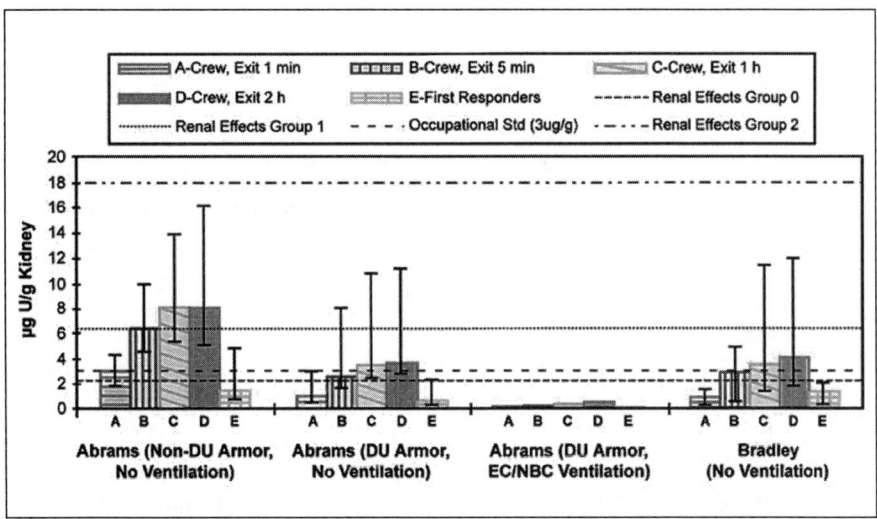

Figure 3.1. HHRA Median Peak Kidney Uranium Concentrations and 10th and 90th Percentiles for All Phases and Scenarios (Compared to REG Levels and the *de facto* Occupational Standard)

by the histograms, and the 90th and 10th percentile values are shown by the upper and lower bars, respectively. The figure also depicts the concentration boundaries of the REG ranges and the *de facto* occupational guideline of 3 µg U/g kidney.

How long DU stays in the body, the organs that it distributes to, and how it is excreted (or eliminated) relates to where it deposits in the respiratory tract and the *in vivo* solubility of the inhaled DU aerosol. Once uranium reaches the kidneys, it is cleared within a few days.

3.5.2 Capstone HHRA Radiological Dose and Risk Results

Internal radiation dosimetry is a complex endeavor. Various scientific and regulatory bodies use an assortment of quantities and units to describe internal radiation doses. The HHRA calculated doses using two currently accepted quantities: the 50-yr committed effective doses, abbreviated as $E(50)$, and the 50-yr committed equivalent doses, abbreviated as $H_T(50)$ where the T stands for the tissue or organ.

Median 50-yr committed effective doses, $E(50)$, for the five scenarios are presented in Table 3.4. The $E(50)$s for the most likely scenarios and upper bound exposures in the Abrams/conventional armor and the Bradley vehicle were ≤5 rem. The $E(50)$ for a 5-min exposure in the unventilated Abrams tank perforated through DU armor was predicted to be 6 rem, and increases to

Table 3.4. Summary of HHRA Median 50-yr Committed Effective Doses by Scenario

Scenarios	E(50), rem			
	Abrams Tank: Conventional Armor, No Ventilation	Abrams Tank: DU Armor, No Ventilation	Abrams Tank: DU Armor, EC/NBC Operating	Bradley Vehicle: Conventional Armor, No Ventilation
Most Likely				
A - Crew, exit in 1 min	2.0	2.2	0.090	0.59
B - Crew, exit in 5 min	3.7	6.0	0.44	1.7
E - First responders	0.92	1.9	0.41	0.89
Upper Bound				
C - Crew, exit in 1 h	4.8	8.3	1.02	2.1
D - Crew, exit in 2 h	5.0	8.7	1.20	2.4

8 rem with a 2-h exposure. The U.S. annual occupational radiation dose limits, as promulgated by the Nuclear Regulatory Commission (NRC), include a 5 rem total effective dose equivalent (10 CFR 20). Although the E(50) is different from the total effective dose equivalent, the concepts are similar, and for radiation protection purposes, the two quantities can be compared (ICRP [60] 1991). The NRC's planned special exposure limit is 10 rem in a year (two times the annual limit, not to exceed five times the annual limit in a lifetime [10 CFR 20.1206]). Although for some scenarios, the E(50)s exceed the occupational radiation limits, all E(50)s are less than the planned special exposure limit. For all scenarios modeled, radiation doses are at levels unlikely to cause adverse health effects.

This table may be useful in bracketing a veteran's E(50) based on the description of events that led to the exposure, but it should not be used to assign individual doses. Biomonitoring, such as urine bioassay, should be used to provide definitive dose estimates, if possible.

Unlike the results of the chemical effects analysis in which the Abrams tank perforated through non-DU armor (Phase I) was highest, the radiological doses were highest from the Abrams tank perforated through DU armor (Phase III). The reason for this difference is that the DU aerosols collected in Phase I were more soluble than the aerosols collected from Phase III. Doses to a crewmember in an Abrams tank with its EC/NBC ventilation system operating when perforated through DU armor were about an order of magnitude below the levels in a perforated Abrams tank in which the ventilation system was not operating. Radiation doses at the levels predicted are unlikely to cause adverse health effects.

Fifty year committed equivalent doses, $H_T(50)$, to various tissues and organs were calculated including the lung, bone surface, kidney, red marrow, and liver (see Attachment 3, Appendix A for a complete list of tissues and organs). The doses to the lung were higher by at least a factor of 10 than the doses to the tissues/organs listed above. The predicted *median* $H_T(50)$s calculated for the lung are summarized in Table 3.5 for each of the scenarios. The $H_T(50)$s for the most likely exposures in these vehicles were between 0.7 and 3.3 rem in a ventilated Abrams tank and between 14 and 44 rem in an unventilated Abrams tank. Doses in an unventilated Bradley vehicle fall between the Abrams ventilated/unventilated scenarios. The upper bounds for exposures ranged from 7.6 to 8.7 rem in a ventilated Abrams tank and 38 to 61 rem in an unventilated Abrams tank. The doses in the Bradley vehicle were between these ranges. The U.S. NRC annual occupational radiation dose limits include a 50-rem committed dose equivalent (10 CFR 20). Although the $H_T(50)$ is different from the committed dose equivalent, the concepts are similar, and for radiation protection purposes, the two quantities can be compared (ICRP [60] 1991). Except for the case in which an Abrams tank was perforated through DU armor and the stay-time was 1 to 2 h, the predicted doses to the organs were less than this occupational limit. For all scenarios modeled, organ doses are at levels unlikely to cause adverse health effects.

The risks for cancer mortality were calculated based on the sum of the individual organs' risks rather than by multiplying the effective dose by an effective dose risk coefficient. Using the organ risk factors approach effectively handles the non-uniformity in dose distribution with the majority of the dose being delivered to the lung, which is a relatively radiosensitive organ. This approach was used because of the availability of "generic" lung cancer mortality risk coefficients based on alpha emitters and the availability of risk coefficients for the other major organ systems (see Attachment 3, Section 6.5). Using summed organ risks resulted in risks that were about 35% higher than the risks based on whole body effective doses. The lifetime cancer

Table 3.5. Summary of HHRA Median 50-yr Committed Equivalent Doses to the Lung by Scenario

Scenario	Lung $H_T(50)$, rem			
	Abrams Tank: Conventional Armor, No Ventilation	Abrams Tank: DU Armor, No Ventilation	Abrams Tank: DU Armor, EC/NBC Operating	Bradley Vehicle: Conventional Armor, No Ventilation
Most Likely				
A – Crew, exit in 1 min	14	18	0.66	5.2
B – Crew, exit in 5 min	32	44	3.3	14
E – First responders	8.8	14	3.1	6.7
Upper Bound				
C – Crew, exit in 1 h	38	60	7.6	20
D – Crew, exit in 2 h	39	61	8.7	22

mortality risks from inhalation of DU aerosols were calculated using the Linear No-Threshold model of effect that may overestimate risks at the levels predicted in this study and are, therefore, thought to be protective of health. Table 3.6 summarizes the median estimated increased lifetime cancer mortality from inhaled, deposited DU aerosols for the five exposure scenarios evaluated.

The median probability of lifetime cancer mortality from DU aerosol inhalation within the most likely exposure scenarios ranged from 0.005 to 0.32%. The upper bound ranged from 0.057 to 0.45%. The risk of 0.12% to a crewmember exiting an unventilated Abrams tank 1 min after perforation through DU armor implies that, on average, the chance over an individual's lifetime of dying of cancer from this exposure is 0.12% greater than the chance of that same individual dying from the natural or background rate of cancer mortality, which is 23.6% for males (Ries et al. 2003). Risks to a crewmember in an Abrams tank (perforated through DU armor) in which the EC/NBC ventilation system was operating were at least a factor of 4 less than the risks in an unventilated Abrams tank (perforated through DU armor).

Figure 3.2 compares the median increased fatal cancer risks by scenario for Level I personnel. In addition to the histogram of the median risks, the figure provides 10th and 90th percentile risks and shows how these levels relate to U.S. occupational dose limits, emergency guidelines, and dose levels used by the U.S. DoD and international community for combat operations (expressed as radiation exposure status or RES categories; see Section 5.1 for details.). Risk lines represented in the chart include the following:

- 0.5 rem, the U.S. annual exposure limit for members of the general public with permission (10 CFR 20).

Table 3.6. Summary of HHRA Median Lifetime Risk of Fatal Cancer from DU Inhalation by Scenario

Scenario	Lifetime Risk Increase of Fatal Cancer (%)			
	Abrams Tank: Conventional Armor, No Ventilation	Abrams Tank: DU Armor, No Ventilation	Abrams Tank: DU Armor, EC/NBC Operating	Bradley Vehicle: Conventional Armor, No Ventilation
Most Likely				
A – Crew, exit in 1 min	0.11	0.12	0.0049	0.034
B – Crew, exit in 5 min	0.20	0.32	0.025	0.099
E – First responders	0.050	0.10	0.023	0.052
Upper Bound				
C – Crew, Exit in 1 h	0.27	0.44	0.057	0.12
D – Crew, Exit in 2 h	0.28	0.45	0.065	0.14

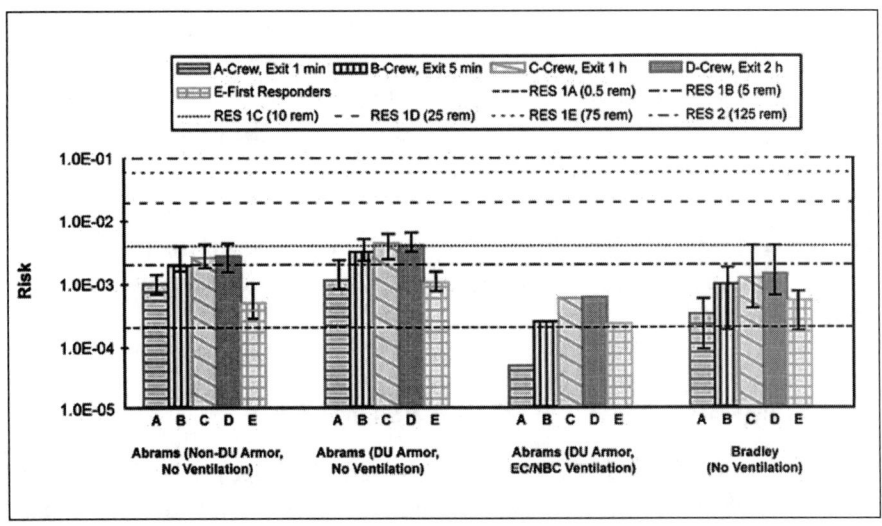

Figure 3.2. HHRA Median Radiation Risks with 10th and 90th Percentile Bars from Perforation of the Vehicles (Compared to Radiation Standards and Emergency Guidelines)

- 5 rem, the U.S. annual exposure limit for workers as well as being the U.S. Environmental Protection Agency's (EPA's) Protective Action Guide (PAG) for all workers in an emergency (EPA 1992).
- 10 rem, the U.S. Nuclear Regulatory Commission's (NRC's) upper annual limit for a planned special exposure and the EPA PAG for protecting valuable property (10 CFR 20.1206, EPA 1992).
- 25 rem, the EPA PAG for life saving actions. Doses higher than 25 rem may be incurred on a voluntary basis to save lives or to protect large populations (EPA 1992).

The figure illustrates that the risks for even the longest stay-times in a ***ventilated*** Abrams tank perforated through DU armor by a DU munition are well below the risks associated with U.S. occupational radiation dose limits. For rapid exits (1 min or less), the risks are slightly greater than the risks associated with the general population dose limit of 0.5 rem. For all vehicle types, the estimated risks at the 90th percentile are below or slightly exceed (by less than 10%) the risks associated with planned special exposures. Finally, the risks for first responders are estimated to be below the risks associated with the occupational limit of 5 rem/yr.

3.5.3 Multiple Perforations

The analyses in this section are based on perforation through a crew compartment of an Abrams tank or a Bradley vehicle from a single large-caliber

DU munition. For cases in which the crew compartment is perforated twice and the crewmembers remain inside during both events with hatches closed, the kidney uranium concentration and radiation doses can be roughly approximated by multiplying by two. The possibility of becoming ill (from transient or protracted kidney effects) increases with vehicle perforations depending on the vehicle configuration. DU aerosol concentrations and predicted intakes, peak kidney uranium concentrations, and E(50)s from the two sets of double shots fired during the Capstone study are discussed in Attachment 3, Chapter 5.

3.6 COMPARISON WITH OTHER ESTIMATES

Prior to this HHRA, USACHPPM (2000) and The Royal Society (2001, 2002) conducted the most extensive studies of Level I intakes and consequent radiological doses and chemical organ concentrations. At the outset of this discussion, it is important to recognize the fundamental difference in the objectives of these studies. The objective of The Royal Society studies was to provide a "central estimate and a worst-case estimate" of the radiological doses and chemical concentrations for Level I personnel in an armored vehicle, while the objective of the USACHPPM report was to estimate reasonable lower and upper bounds for Level I exposures, including intake, radiological dose, and kidney uranium concentrations for personnel in armored Abrams tanks and Bradley vehicles.

The Royal Society (2001, 2002) and USACHPPM (2000) considered intakes from inhalation and secondary ingestion (hand-to-mouth transfer of DU) as well as the small contribution from external radiation. In general, standard ICRP biokinetic models were used in these studies, and similar assumptions were made regarding the aerosol characteristics (solubility, particle-size distribution) and physiological parameters required to estimate radiation dose and kidney concentrations. The primary differences in their approaches was how initial DU aerosol concentrations were estimated and how the uncertainty in each of these parameters, particularly the uncertainty in aerosol concentrations, was handled.

The Royal Society (2001, 2002) provided two estimates of radiological doses and kidney concentrations. The first was their "central estimate," which was defined as:

> "...intended to be a central, representative value, using likely values of parameters according to the information available, and where information is lacking, values that are unlikely to underestimate exposures greatly. The central estimate is intended to be representative of the

average for the group (or population) of people exposed in that situation. However, the term average is not used, because the central estimates are not based on statistical analyses of data. It is recognized that for individuals in each group values could be greater than (or less than) the central estimate."

The Royal Society (2001, 2002) also estimated worst-case radiation doses and kidney concentrations where worst case is defined as an estimate:

"…which uses values at the upper end of the likely range, but not extreme theoretical possibilities. The aim is that it is unlikely that the value for any individual would exceed the worst-case. Thus the worst-case should not be applied to the whole group to estimate, for example, the number of excess cancers that might be induced. One aim of the worst-case assessments is to try to prioritise further investigation. If even the worst-case assessment for a scenario leads to small exposures, then there is little need to investigate it more closely. If, however, the worst-case assessment for a scenario leads to significant exposures, it does not necessarily mean that such high exposures have occurred, or are likely to occur in a future battlefield, but that further information is needed."

For each parameter, USACHPPM (2000) developed a most probable value, and then, using available data and professional judgment, they developed reasonable estimates of the lowest reasonable values and estimates of the distribution function between these values. USACHPPM then used a Monte Carlo simulation that generated a probabilistic distribution of intakes based on the defined input parameters. Distributions of radiological doses and kidney concentrations were generated using the intake distribution and the distributions of the parameters required to convert intakes into doses and organ concentrations. Figure 3.3 illustrates the results of this simulation for calculating the committed effective dose equivalent (CEDE) for their upper bound estimate for inhalation exposure (discussed below). Summary results were reported in terms of median dose.

The second difference between the assessments made by USACHPPM and The Royal Society was the method used to estimate the aerosol concentrations inside the vehicle at the time of perforation. Fliszar et al. (1989) provided aerosol concentration data gathered inside the vehicle during perforation. The data were obtained from aerosol samplers placed inside the vehicle prior to perforation. Because of experimental difficulties described in Fliszar et al. (1989) and summarized in USACHPPM (2000), there was significant uncertainty in run durations for air samplers that were not damaged when the vehicle was perforated. USACHPPM addressed this uncertainty by

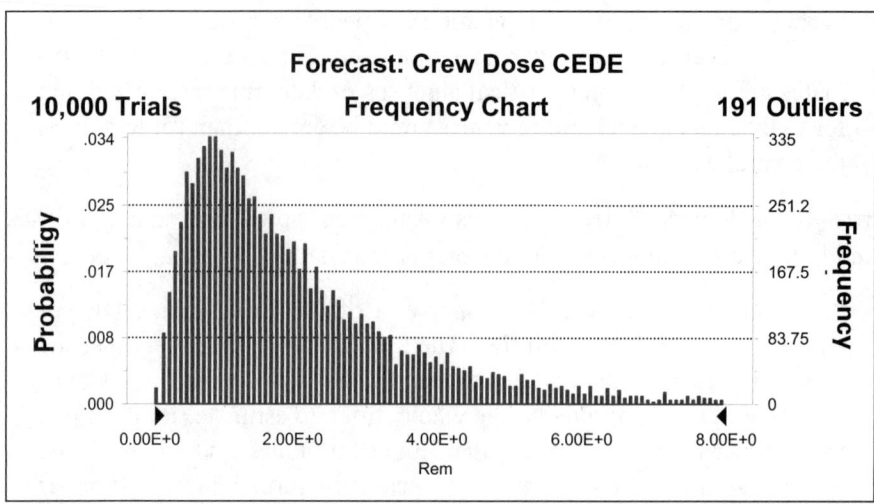

Figure 3.3. Estimates of Crewmember Committed Effective Dose Equivalent Probability Distribution for the Upper Bound Calculation from USACHPPM (2000)

establishing an upper and a lower estimate of the air sampler run durations to establish an upper and a lower bound estimate of the aerosol concentration at the time of perforation. Figure 3.3 shows the distribution of doses from the upper bound assessment.

The Royal Society opted to use published estimates of the fraction of a penetrator aerosolized during perforation and the mass of the penetrator coupled with an estimate of the volume of the crew compartment (12 m^3) to estimate the initial aerosol concentration. Their central estimate was based on the assumption that 2 to 3% of a 4-kg penetrator (approximately 100 g) was aerosolized and dispersed in the 12-m^3 volume. The worst-case estimate was based upon the assumption that 20% of a 5-kg penetrator was aerosolized (approximately 1000 g). This illustrates The Royal Society's process for its worst-case estimate—it used upper estimates for all parameters in the estimate. In this case, both the percent aerosolized in the crew compartment and the mass of the penetrator were increased.

USACHPPM used a distribution of stay-times for the crew after perforation that ranged from a lower bound of 10 sec to an upper bound of 120 sec, with a most likely value of 60 sec. The Royal Society used a value of 60 sec as its central estimate. A comparison of the results of these two analyses with the HHRA, **assuming no ventilation,** is shown in Table 3.7.

Considering the variety of approaches used in these estimates, there is good agreement with the risk assessments done for penetrations of the Abrams

Table 3.7. Comparison of Median Doses and Concentration Values

Assessment Parameter	USACHPPM 2000[a] (Median Values)		Capstone Health Risk Assessment Scenario A[b], No Ventilation (Median Values)			The Royal Society (Central Estimate)[c]
	Lower	Upper	Abrams: Non-DU Armor	Abrams: DU Armor	Bradley Vehicle	
E(50) (rem)	0.179	1.59	1.97	2.16	0.589	2.2
Lung H_T(50) (rem)	1.49	13.2	14.3	17.5	5.21	17.8
Peak kidney U conc. (µg U/g kidney)	0.17	1.46	2.99	1.08	0.988	4

(a) Mean exposure duration was approximately 1 min.
(b) Scenario A is a 1-min exposure.
(c) The Royal Society assumed a 1-min exposure.

tank, which is representative of a heavily armored vehicle. These results also compare well to the E(50) of 4 rem and 3 µg U/g kidney uranium concentration estimates reported by Fetter and von Hipple (1999).

For its worst-case estimate, The Royal Society assumed a stay-time of 41 min. Based on data presented by The Royal Society (2001), it was assumed that the concentration in the vehicle remained constant for 1 min and decreased by a factor of 10 every 10 min out to 41 min. There is no directly comparable time-concentration scenario for the USACHPPM report. Scenario C of the Capstone report was chosen for comparison because the stay-time (60 min) was the closest to the 41 min used by The Royal Society. Table 3.8 below compares the Capstone study Scenario C median and 90th percentile results (**assuming no ventilation**) with the worst case described by The Royal Society.

Table 3.8. Comparison of Capstone HHRA Median and 90th Percentile 1-h Dose and Concentration Estimates with Worst-Case Estimates Made by The Royal Society

Assessment Parameter		Capstone Health Risk Assessment Using Scenario C (1-h Stay-Time, No Ventilation)			The Royal Society: Worst-Case
		Abrams: Non-DU Armor	Abrams: DU Armor	Bradley Vehicle	
E(50) (rem)	Median	4.85	8.33	2.08	110
	90th Percentile	8.56	16.4	6.84	
Lung H_T(50) (rem)	Median	35.7	58.5	16.0	950
	90th Percentile	59.0	85.6	53.2	
Peak kidney U concentration (µg U/g kidney)	Median	8.17	3.55	3.53	400
	90th Percentile	14.0	10.8	11.5	

In general, The Royal Society's worst-case estimates are more than an order of magnitude greater than those derived from the Capstone data. Because of the use of worst-case assumptions used in The Royal Society reports for nearly every parameter, the larger estimated doses and kidney uranium concentrations are not surprising.

3.7 REFERENCES

10 CFR 20. Code of Federal Regulations, Title 10, *Energy*, Part 20, "Standards for Protection Against Radiation—Planned Special Exposures."

Agency for Toxic Substances and Disease Registry (ATSDR). 1999. *Toxicological Profile for Uranium.* Report TP 90-29, Atlanta, Georgia.

Deployment Health Support Directorate (formerly OSAGWI), US Department of Defense. 2004. Letter from COL D. Sulka, Director, Force Health Protection, to LTC MA Melanson, US Army Center for Health Promotion and Preventive Medicine, April 22, 2004.

Fetter S and FN von Hippel. 1999. "The Hazard Posed by Depleted Uranium Munitions." *Science & Global Security* 8(2):125-161.

Fliszar RW, EF Wilsey, and EW Bloore. 1989. *Radiological Contamination from Impacted Abrams Heavy Armor*. Technical Report BRL-TR-3068, Aberdeen Proving Ground, Ballistic Research Laboratory, Aberdeen, Maryland.

Fulco CE, CT Liverman, and HC Sox, eds. 2000. "Depleted Uranium." In *Gulf War and Health*, Vol. 1, Chapter 4, National Academy Press, Washington, DC.

Guilmette RA, MA Parkhurst, G Miller, FF Hahn, LE Roszell, EG Daxon, TT Little, JJ Whicker, YS Cheng, RJ Traub, GM Lodde, F Szrom, DE Bihl, KL Creek, and CB McKee. 2004. *Human Health Risk Assessment of Capstone Depleted Uranium Aerosols*. PNWD-3442, prepared for the US Army by Battelle under Chemical and Biological Defense Information Analysis Center Task 241, DO 0189, Aberdeen, Maryland.

International Commission on Radiological Protection (ICRP). 1975. *Report of the Task Group on Reference Man.* ICRP Publication 23, Pergamon Press, Oxford, United Kingdom.

International Commission on Radiological Protectection (ICRP). 1979. *Limits for Intakes of Radionuclides by Workers*. ICRP Publication 30, Part 1, Pergamon Press, Oxford, United Kingdom.

International Commission on Radiological Protection (ICRP). 1991. *1990 Recommendations of the International Commission on Radiological Protection*. ICRP Publication 60, Pergamon Press, Oxford, United Kingdom.

International Commission on Radiological Protection (ICRP). 1994a. *Human Respiratory Tract Model for Radiological Protection*. ICRP Publication 66, Pergamon Press, Oxford, United Kingdom.

International Commission on Radiological Protection (ICRP). 1994b. *Dose Coefficients for Intakes of Radionuclides by Workers*. ICRP Publication 68, Pergamon Press, Oxford, United Kingdom.

International Commission on Radiological Protection (ICRP). 1995a. *Basic Anatomical and Physiological Data for Use in Radiological Protection: the Skeleton*. ICRP Publication 70, Pergamon Press, Oxford, United Kingdom.

International Commission on Radiological Protection (ICRP). 1995b. *Age-Dependent Doses to Members of the Public from Intake of Radionuclides: Part 3 Ingestion Dose Coefficients*. ICRP Publication 69, Pergamon Press, Oxford, United Kingdom.

International Commission on Radiological Protection (ICRP). 2003. *Basic Anatomical and Physiological Data for Use in Radiological Protection: Reference Values*. ICRP Publication 89, Pergamon Press, Oxford, United Kingdom.

Parkhurst MA, F Szrom, RA Guilmette, TD Holmes, YS Cheng, JL Kenoyer, JW Collins, TE Sanderson, RW Fliszar, K Gold, JC Beckman, and JA Long. 2004. *Capstone Depleted Uranium Aerosols: Generation and Characterization, Volumes 1 and 2*. PNNL-14168, prepared for the U.S. Army by Pacific Northwest National Laboratory, Richland, Washington.

Ries LAG, MP Eisner, CL Kosary, BF Hankey, BA Miller, L Clegg, A Mariotto, MP Fay, EJ Feuer, and BK Edwards (eds). 2003. SEER Cancer Statistics Review, 1975–2000, National Cancer Institute. Bethesda, Maryland. Accessed online in January 2004 at http://seer.cancer.gov/csr/1975_2000,2003.

The Royal Society. 2001. *The Health Hazards of Depleted Uranium Munitions Part I*. Policy Document 6/01, London, United Kingdom. Online report available at URL: www.royalsoc.ac.uk in the Science Policy Section.

The Royal Society. 2002. *The Health Hazards of Depleted Uranium Munitions Part II*. Policy Document 5/02, London, United Kingdom. Online report available at URL: www.royalsoc.ac.uk in the Science Policy Section.

US Army Center for Health Promotion and Preventive Medicine (USACHPPM). 2000. *Depleted Uranium—Human Exposure Assessment and Health Risk Characterization in Support of the Environmental Exposure Report "Depleted Uranium in the Gulf" of the Office of the Special Assistant to the Secretary of Defense for Gulf War Illnesses, Medical Readiness and Military Deployments (OSAGWI), OSAGWI Levels I, II and III Scenarios, 15 September 2000*. Health Risk Assessment Consultation No. 26-MF-7555-00D, Aberdeen Proving Ground, Maryland. Online report available at URL: in the Environmental Exposure Reports Section.

US Environmental Protection Agency (EPA). 1992. *Manual of Protective Action Guides and Protective Actions for Nuclear Incidents*, EPA 400-R-92-001, Office of Radiation Programs, Washington, DC.

4.0 Level II and Level III Exposures

Exposure Levels II and III from Operation Desert Storm (ODS) were defined by OSAGWI (1998, 2000) and USACHPPM (2000) to distinguish the longer depleted uranium (DU) exposure of personnel engaged in certain recovery, repair, and salvage activities from those whose entry and stay in a DU-contaminated vehicle was relatively short. As applied to this analysis, exposure Levels II and III from ODS are defined as follows:

- Level II includes military personnel and a small number of DoD civilian employees whose job functions required them to work in and around vehicles containing DU fragments and particles. These individuals were not in the vehicle at the time of impact and did not immediately enter the vehicle after it was struck. This group performed a variety of tasks, such as battle damage assessment, repairs, explosive ordnance disposal, and intelligence gathering. They typically entered vehicles well after the initial suspended aerosol had dissipated or settled onto interior surfaces. They may have inhaled DU residues that were resuspended by their physical activities, ingested DU through hand-to-mouth transfer, or spread contamination on their clothing. DoD personnel who were involved in cleaning up DU residues generated during other events, such as the July 11, 1991, explosion and fires at the Camp Doha North Compound, are also included in this group.

- Level III is an "all others" group whose exposures were brief or incidental. This group includes personnel who entered DU-contaminated Iraqi equipment, were downwind from burning Iraqi or U.S. equipment struck by DU rounds, or were downwind from burning DU ammunition (e.g., personnel at Camp Doha during the July 11, 1991, explosions and fire). While these individuals could have inhaled airborne DU particles, the possibility of receiving an intake high enough to cause health effects is unlikely.

These categories would also apply to Operation Iraqi Freedom and to future combat operations.

The primary difference between Level I exposure scenarios and the Level II and III scenarios is the time difference between vehicle perforation and the initiation of personnel activities that may lead to DU intakes. Level II and III personnel would not have been in, on, or near the vehicles at the time of perforation. Rather they would have entered the vehicles at an extended time after perforation. The DU aerosol concentration inside a perforated vehicle decreases rapidly post-impact (Parkhurst et al. 2004, Attachment 1). Level II and Level III personnel are potentially exposed by inhalation of DU that has been resuspended from surfaces as a result of their physical activities in and on struck vehicles or by ingestion (hand-to-mouth transfer) of DU residues that have been transferred from surfaces to hands.

Post perforation, the DU aerosols will settle on surfaces inside and outside of the vehicle and may be inhaled if the material is resuspended into the air. The physical activities of the personnel in and around a damaged vehicle, as well as the condition of the vehicle and the ambient environment, are key factors that may lead to resuspension of DU from surfaces into the air. Removable DU contamination can be found on the surfaces of a vehicle damaged with DU munitions or on a vehicle that has sustained damage to or a breach of its DU armor package. More removable DU may be found on the inside surfaces than on the exterior surfaces of a perforated vehicle (Attachment 1). DU deposited on surfaces may be transferred to an individual's hands, and then may be ingested through hand-to-mouth transfer. Therefore, inhalation of resuspended DU and hand-to-mouth ingestion of DU are the primary exposure pathways for Level II and III personnel.

The primary exposure pathways are the same for Level II and III personnel; however, the time spent by personnel inside the vehicles is different between the two levels. The physical activities performed in and around the vehicles may also be different for Level II and III personnel. Level II personnel spend more time inside vehicles because their jobs require them to work in and around damaged vehicles. Typical Level II jobs involve entering damaged vehicles to repair them, gather intelligence, assess battle damage, or dispose of explosive ordinance. Level III personnel enter damaged vehicles because of curiosity rather than mission requirements. OSAGWI (2000) and USACHPPM (2000) provide additional details describing the Level II and III categories including a detailed breakdown of the specific military specialties and job titles that are in the Level II category.

The data required for portions of the Level II and III assessments were available prior to the Capstone DU Aerosol Study (Attachment 1 and 2). As summarized in USACHPPM (2000) and OSAGWI (2000), a number of tests have been performed to characterize the aerosols generated when DU munitions burn and those generated outside armored vehicles when they were struck by DU munitions. A summary of the available data and assessments of the quality of the data, as well as gaps in the data, was published by USACHPPM (2000). Two data gaps, particular to Level II and Level III exposures, include measurements of DU resuspended into the air and measurements of DU hand contamination. The Capstone DU Aerosol Study report (Attachments 1 and 2) has provided data that can be used to estimate DU intakes for Level II and Level III personnel from the inhalation of resuspended DU and the ingestion of hand contamination. This chapter discusses the approach used to estimate the Level II and III inhalation and ingestion exposures from the Capstone data. This discussion is followed by brief summaries of exposures outside the vehicles and from the smoke plume resulting from burning DU munitions as discussed in USACHPPM (2000). This chapter concludes with a comparison of the Capstone, USACHPPM (2000), and The Royal Society (2001, 2002) estimated exposures.

4.1 METHODS

This section provides a summary of the approaches used to estimate the Level II and Level III intake rates, E(50) rates, and peak kidney uranium concentration rates from the inhalation of resuspended DU inside the vehicle as well as the ingestion that potentially occurs from hand-to-mouth transfer (also called secondary or incidental ingestion) of DU residues deposited on surfaces of perforated vehicles. The details of the calculations are contained in Szrom et al. (2004), which is Attachment 4 of this report. An important underlying assumption for these analyses is that the samples used from the Capstone field tests are representative of vehicle conditions that Level II and Level III personnel would encounter. An additional assumption is that the physical activities performed by Capstone sample recovery personnel are surrogates for the types of physical activities performed by Level II and Level III personnel in and around DU contaminated vehicles. The methods used to assess the Level II exposures from clean-up operations, such as the Camp Doha fire, are documented in USACHPPM (2000) and are not repeated here.

4.1.1 Inhalation

Two Capstone study datasets were used to estimate DU inhalation intakes and subsequent radiological doses and peak kidney uranium concentrations for Level II and Level III personnel. The first dataset evaluated included the personal cascade impactors (P-CIs) worn by sample recovery personnel. Some of the sample recovery personnel were performing activities that were similar to Level II personnel working in DU contaminated vehicles. These sample recovery personnel worked inside the vehicle removing samples, reloading samples, and taking equipment in and out of the vehicle. The P-CIs were operated while sample recovery activities were performed in and around the target vehicles. The data from these air samples were used as a basis to estimate DU aerosol concentrations that could lead to inhalation intakes by Level II or III personnel. Personnel who entered the vehicles after Phase I, Phase II, and Phase III shots performed sample recovery operations and wipe-tests surveys. In addition to sample recovery and wipe-test surveys, some of the personnel who were monitored in Phase IV performed the functions of Explosive Ordnance Personnel and Battle Damage Assessment and Recovery personnel. As discussed in Attachment 4, the P-CIs were turned on when Capstone personnel entered the contaminated zone in the test facility and were turned off when they exited. Consequently, the P-CIs measured the DU concentration as the monitored personnel approached, entered, worked inside, exited, and walked away from the vehicle when their task was completed.

The second dataset evaluated included the cascade impactors (CIs) from two sampling arrays that collected aerosols just prior to the initiation of sample recovery operations and during sample recovery. This dataset provides estimates of DU aerosol concentrations inside the vehicle before, during, and after physical activities were performed in the vehicle. During Capstone Study Phase I Shots 6 and 7, the sampler array at the loader's position was designated for resuspension studies and was not run until several hours after vehicle perforation. In both cases, a residual aerosol sample was run before any recovery personnel entered the vehicle. This sampling began about 2-1/2 h after PI-6 and about 3-1/2 h after PI-7 and operated for 20 min. The next four samples from PI-6 operated in conjunction with personnel recovery activities; the last four samples ran after all personnel had exited the vehicle. Following the PI-7 shot residual sampling, two 20-min samples were taken. The first of these two samplers began operating as personnel entered the vehicle. The second sampler operated while personnel continued their activities in the vehicle and ended as soon as all personnel exited the vehicle. A final sampler began operating as soon as all personnel were out of the vehicle and continued for approximately 10 min. The CI data from these arrays were used to estimate intake rates, E(50) rates, and peak kidney uranium concentration rates.

CI data were used in the analysis of intake and dose rates for both the breathing-zone monitors (P-CIs) and the area-monitoring arrays (array CIs) so that particle size, which affects respirability, could be considered in the calculations. The concentrations calculated for the breathing zone monitors and the area monitoring arrays were adjusted to compensate for wall loss. The DU aerosol particle size distribution was represented by nine monodisperse aerosols, rather than fitting unimodal or bimodal distributions to the CI data to describe the activity median aerodynamic diameters (AMADs) of the sampled aerosols. Stage-specific DU concentrations were multiplied by the breathing rate for a reference worker (1.2 m^3/h), and the total length of time the CI was activated to estimate the inhalation intake. For the P-CIs, the time interval included the time the individual spent inside the vehicle and the time spent engaged in sample collection activities around the vehicle. Lung fluid dissolution functions derived from the Capstone cyclone aerosol samples were selected by particle size and vehicle type. The Integrated Modules for Bioassay Analysis–Uranium (IMBA-URAN) code (James 2002) was used to calculate peak kidney uranium concentrations and organ-equivalent dose conversion factors. (The details of this calculation are contained in Attachment 4.)

The stage-specific intakes estimated from each of the eight stages and the backup filter of each CI were multiplied by the appropriate dose conversion factors (DCFs). The peak kidney uranium concentrations and organ doses for the nine stages were summed, and the 50-yr committed effective dose, E(50), was calculated for each CI. The results were then divided by the exposure time to calculate rates. In both cases, these results provide a good estimate of the intake and dose rates by vehicle type for Level II personnel but probably represent an upper bound estimate for Level III personnel, because all measurements were collected from personnel performing Level II-like actions. The mean of the intake rates, E(50) rates, 50-yr committed equivalent doses, H_{Lung}(50) rates, and 24-h (peak) kidney uranium concentration rates for the P-CIs are presented in Table 4.1. The CI-array area monitor mean data are

Table 4.1. P-CI Summary Results for Level II Personnel Exposures In and Around Vehicles

Parameter	DU Intake (mg/h)	E(50) (rem/h)	H_{Lung}(50) (rem/h)	24-h Kidney U Conc. (µg U/g-h)
Mean	0.447	1.97E-03	1.23E-02	2.89E-03
Standard Deviation	0.358	2.36E-03	1.64E-02	2.47E-03

presented in Table 4.2. These tables also contain calculations of the standard deviations for these data.

Table 4.2. CI-Array Summary Rate Results for Level II Area Monitors Inside Vehicles

Parameter	DU Intake (mg/h)	E(50) (rem/h)	$H_{Lung}(50)$ (rem/h)	24-h Kidney U Conc. (µg U/g-h)
Mean	14.5	7.80E-02	5.59E-01	1.43E-01
Standard Deviation	11.2	7.23E-02	5.37E-01	1.22E-01

As shown in Tables 4.1 and 4.2, all of the values from the "resuspension" area monitor arrays were higher than the P-CI results. This difference between the two sampler results is related to 1) the sampler location with regard to the person's breathing zone, and 2) the sample collection location with respect to the vehicle. The sampling intervals for the P-CIs included time on or outside the vehicle as well as the time spent inside the vehicle. The CI arrays, on the other hand, operated entirely inside the vehicle beneath the loader position hatch, which was used for entry and exit by the sample recovery team.

The P-CI data covers each of the shots in the four phases and may be more representative of potential Level II and Level III personnel exposures because they collected aerosol within the breathing zone and they were used to incorporate a combination of activities both inside and outside the vehicle. However, they may underestimate the dose rates when applied to exposures entirely within the vehicle. The CI-array area monitors can be used to estimate potential intakes and doses from in-vehicle resuspension. However, these results are based on only two shots in Phase I, and because they were not taken within the breathing zone inside the vehicles, may overestimate dose rates.

The approach recommended for estimating intake, E(50), and peak kidney uranium concentrations from inhalation of DU aerosols is based on knowledge regarding fraction of time spent inside a vehicle versus the total time on or around the vehicle. The following criteria can be used to determine the approach to be used.

- If the total time in the vicinity of the DU-perforated vehicle is known and the fraction of time inside the vehicle can be estimated, it is recommended that the CI-array area monitor data be applied for the time spent working inside the vehicle and the P-CIs be used for the time working on and around the vehicle.

- If only the total time spent in, on, or around the perforated vehicle is known, it is recommended that the intakes be estimated as a range of values with the P-CI values representing the most likely value and the area monitoring data representing the upper bound. It is recognized that the actual dose could be less than the most likely estimate. A more detailed explanation is contained in Attachment 4.

4.1.2 Ingestion

Estimates of potential Level II and Level III DU ingestion from hand contamination were made using the cotton glove contamination data collected during the Capstone study (Attachment 1). To estimate hand contamination, the amount of DU on the cotton gloves worn by personnel working in and around DU contaminated vehicles was measured and then adjusted for the estimated difference in DU-adherence to skin versus cotton gloves. Estimates of hand-to-mouth transfer were calculated from the fraction of the hand contamination that potentially could be ingested. These intake estimates were used as input to dose modeling, resulting in rate calculations of $E(50)$, $H_{Lung}(50)$, $H_{Bone\ Surface}(50)$, and the 24-h (peak) kidney uranium concentration.

Various Capstone sample recovery team members wore cotton inspection gloves over personal protective equipment (PPE) work gloves. The types of duties performed by sample recovery personnel included entering and exiting the vehicles, removing air samplers and deposition trays, performing instrumentation surveys, and collecting wipe-test samples from pre-marked surface areas of the vehicles. A total of 28 pairs of gloves were collected from the following Capstone field tests: PI-1, PI-2, PI-5, PIII-2, PIV-1, PIV-3, and PIV-4. The test phases and shots are identified by the phase (PI, PII, PIII, and PIV) and the shot (1 to 7). (Due to a logistical oversight, no glove samples were collected from the Phase II [Bradley vehicle] tests.) Information was recorded about the sample recovery duties performed by the individuals wearing the gloves, as well as the time duration for which the gloves were worn while performing these sample recovery duties. The descriptions of the duties as well as direct observation of the individuals performing these functions were used to categorize the glove data sets as either Level II- or Level III-like activities. After sample recovery personnel performed their duties, the gloves were collected and subsequently analyzed for U-238, as detailed in Attachment 2, Appendix A, Section A.2.2.

To estimate an ingestion dose rate from the Capstone glove contamination data, several important assumptions must be made. Two of these assumptions include the quantity of DU that would get on the individual's hands (compared to what was measured on cotton gloves) and the quantity of the hand contamination that is subsequently ingested. Distributions describing the various input parameters were modeled as probability distributions, after which Monte Carlo analyses were performed to develop a distribution of potential ingestion. The Monte Carlo analyses using Crystal Ball (software product of Decision Engineering, Denver Colorado) calculated a distribution (from 10,000 trials) of ingestion intake rates (and subsequent dose rates and peak kidney uranium concentration rates (using the IMBA-URAN DCFs) using the following equation:

$$\text{Intake rate} = \text{DU "on glove" rate} \times \text{skin/glove "on-take" fraction} \times \text{fraction of DU on hands (skin) ingested}.$$

The input parameters were modeled using actual data distributions (DU "on glove" rates) where possible. Where actual data were not available, triangular distributions with defined minimum, most-likely, and maximum values were used for the remainder of the parameters in the equation above. Triangular distributions were constructed by Crystal Ball using the most-likely value to define the apex of the triangle and the minimum and maximum values to define the length of the base. The triangle is constructed so that the total probability (the area of the triangle) is equal to one, and the probability of the maximum or minimum occurring is zero.

DU "on glove" rates. The "on gloves" mass and "on gloves" mass rates calculated for each pair of gloves were detailed in Attachment 4, Table 21. The table is sorted by exposure Level II (highlighted) and III (not highlighted). For Level II estimates, the nineteen Level II DU "on glove" mass rates were modeled with each value having the same relative probability (1/19). For Level III estimates, the nine Level III DU "on glove" mass rates were modeled with each value having the same relative probability (1/9).

Skin/glove on-take fraction. Cotton gloves were used as a surrogate for skin/hands. More contaminant adheres to cotton gloves than would adhere to skin because penetration into the fabric can occur. As a result, cotton on-take typically results in an overestimate of the material that would be taken on to the hand (Pierce 1998). Also, the larger glove surface area, when worn over the hand, increases the adherent surface area (Brouwer 1999). A study with a powder contaminant reported ". . . an approximately 70 times higher adherence at the glove . . ." than the skin (Brouwer 1999). Other studies with pesticides have shown factors of 2.4 (Fenske 1989) and 5 (Davies 1983) times higher on cotton gloves than skin. The referenced values were used as a guide to model a distribution for the skin/glove on-take fraction. This is the fraction of the material that adhered to the glove and would be expected to adhere to the hand if the hands were bare and the cotton gloves were not worn. This skin/glove on-take fraction was modeled as a triangular distribution with a minimum value of 0.02 (using 1/50 rather than 1/70), a likeliest value of 0.2 (1/5) and a maximum value of 0.5 (1/2).

Fraction of DU on hands (skin) ingested. The final parameter modeled for the Monte Carlo analysis was the fraction of DU on hands that is ingested. This parameter was also modeled as a triangular distribution with a minimum value of zero and a likeliest value of 0.1 (fingertips are approximately 10% of the surface area of the hand and most likely to be placed in/near mouth) and a maximum value of one. Figure 7 shows the triangular distribution that was used for the fraction of the DU on the hands that was ingested. The minimum value used was "0" in which none of the DU on the hand was ingested. The maximum used was 100% of the DU on the hand was ingested. The most likely value was 10% of the DU on the hand was ingested. As shown in Figure 4.1, the mean value using this distribution was 37%.

IMBA-URAN was used to calculate DCFs for a unit (mg) ingestion of typical U.S. DoD DU, assuming an f_1 (gut absorption fraction) equal to 0.02. For uranium compounds, the ICRP 68 (1994) f_1 defaults for unspecified compounds and most tetravalent compounds are 0.02 and 0.002, respectively. The value 0.02 was used as a conservative assumption. A summary of the ingestion DCFs used is presented in Table 4.3. The bone surface equivalent

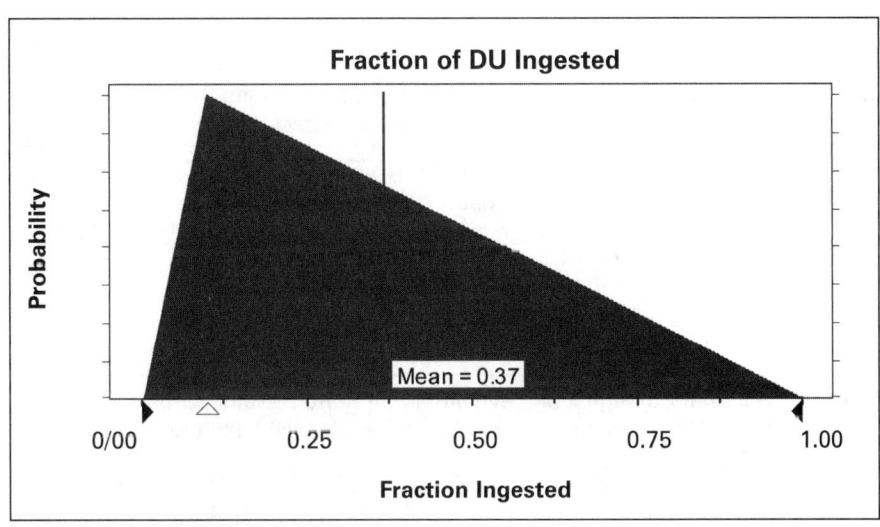

Figure 4.1. Probability Distribution of the Fraction of DU on the Hand Actually Ingested

Table 4.3. Summary of IMBA-URAN Dose Conversion Factors Used

Parameter	DCF
E(50) ingestion DCF (mSv/mg)	6.70E-04
Peak (24-h) kidney U conc. DCF (μg U/g-kidney tissue per mg intake)	7.27E-03
$H_{LUNG}(50)$ (Sv/mg)	3.47E-07
$H_{Bone\ Surface}(50)$ (Sv/mg)	9.90E-06

dose rate DCF is presented for these ingestion scenarios because the bone surface is the organ/tissue with the highest equivalent dose per unit intake via ingestion. IMBA-URAN input and output screens for a unit (1 mg) intake of DoD DU are presented in Attachment 4.

Estimates of the Level II and Level III intake rates, dose rates as E(50) and $H_T(50)$ for bone surface (BS) and lung, and peak kidney uranium concentration rates from ingestion intakes of DU, from the Monte Carlo analyses performed with the Capstone DU glove data, are presented in Tables 4.4 and 4.5. Exposure times were not modeled in the Monte Carlo analyses because of the difficulty in estimating in-vehicle stay-times for all Level II and Level III potential exposure scenarios; therefore, the results in Tables 4.4 and 4.5 are presented as rates.

Due to a logistical oversight, glove data were not available for Phase II. Attachment 4 provides a detailed discussion of how the wipe-test data were

Table 4.4. Summary of Ingestion Level II Monte Carlo Results Based on Mean Glove Contamination (Combined Phases I, III, and IV)

Parameter	Intake Rate (mg/h)	E(50) Rate (rem/h)	$H_{Lung}(50)$ Rate (rem/h)	$H_{BS}(50)$ Rate (rem/h)	Peak Kidney U Conc. Rate (μg/g-h)
Mean	1.06E+01	7.07E-04	3.66E-04	1.05E-02	7.67E-02
Standard Deviation	1.54E+01	1.03E-03	5.36E-04	1.53E-02	1.12E-01

Table 4.5. Summary of Ingestion Level III Monte Carlo Results Based on Mean Glove Contamination (Combined Phases I, III, and IV)

Parameter	Intake Rate (mg/h)	E(50) Rate (rem/h)	$H_{Lung}(50)$ Rate (rem/h)	$H_{BS}(50)$ Rate (rem/h)	Peak Kidney U Conc. Rate (μg/g-h)
Mean	1.78E+00	1.20E-04	6.19E-05	1.77E-03	1.30E-02
Standard Deviation	2.18E+00	1.46E-04	7.58E-05	2.16E-03	1.59E-02

used to provide an estimate of the Phase II ingestion intake rates and subsequent dose rates. In general, there were six sets of PIV-4 glove contamination data that were paired with Phase IV-4 interior wipe-test survey data. These paired data sets were used to calculate surface contamination to glove transfer factors and rates. This analysis showed that, within the limitations of the data, potential ingestion rates could be estimated from wipe test data.

4.1.3 Summary of Inhalation and Ingestion for Level II and Level III Exposures from the Capstone Data

A general strategy for applying the inhalation results is presented in Table 4.6. The data did not allow a distinction to be made between Level II and Level III personnel because the activities performed when the samples were collected were all Level III-like activities. If the total time in the vicinity of the vehicle is known and the fraction of time inside the vehicle can be estimated, it is recommended that the area monitor dose rates, $D_r(AM)$, be used for the spent time inside (T_I) the vehicle and the breathing zone monitor dose rates, $D_r(BZM)$, be used for the time spent approaching the vehicle and on the vehicle or the time spent outside the vehicle (T_O). This will overestimate the inhalation intake values and subsequent dose estimates because the breathing zone values include time spent inside the vehicle. If only the total time is known, it is recommended that the intakes be estimated as a range of values with the breathing zone data representing the lower bound and the area monitoring data representing the upper bound. It is recognized that the actual dose could be less than the lower bound estimate.

A different approach is recommended for estimates of potential intake rates, E(50) rates, and peak kidney uranium concentrations from the ingestion of DU residues. In this case, a distinction was made by separating glove samples into Level III-like or Level III-like personnel exposures based on the activities performed by sample recovery personnel. Table 4.7 summarizes the intake rates, E(50) rates, and peak kidney uranium concentration rates

Table 4.6. Approach for Estimating Potential Inhalation Doses

Option	Dose Rate Application	Calculation
Total exposure time known **and** Time in perforated vehicle known	Use AM for time in vehicle Use BZM for rest of exposure time	$D = T_O D_r(BZM) + T_I D_r(AM)$
Total exposure time known **but** Time in perforated vehicle unknown	**Lower Bound Estimate:** Use BZM for total time **Upper Bound Estimate:** Use AM for total time	**Lower Bound Estimate:** $D_{LB} = T_O D_r(BZM)$ **Upper Bound Estimate:** $D_{UP} = T_I D_r(AM)$

Table 4.7. Summary of the Estimated E(50) and Peak Kidney Uranium Concentration Rates from Inhalation and Ingestion for Level II and Level III Personnel Exposures

Parameter		Inhalation (Levels II and III)		Ingestion (Hand-to-Mouth)	
		Breathing Zone Monitors (BZM)	Area Monitors (AM)	Level II	Level III
DU Intake (mg/h)	Mean	0.447	14.5	10.6	1.78
	Std. Dev.	0.358	11.2	15.4	2.18
E(50) (rem/h)	Mean	1.97E-03	7.80E-02	7.07E-04	1.20E-04
	Std. Dev.	2.36E-03	7.23E-02	1.03E-03	1.46E-04
Peak Kidney U Conc. (µg U/g-h)	Mean	2.89E-03	1.43E-01	7.67E-02	1.30E-02
	Std. Dev.	2.47E-03	1.22E-01	1.12E-01	1.59E-02

derived from the Capstone data for Levels II and Level III personnel. Estimates of dose and peak kidney uranium concentration can be obtained by multiplying the mean rates shown in the table by estimates of exposure duration for a single acute exposure. The exposure duration (time) used for ingestion data should be the time personnel were in contact with DU residues deposited on surfaces. Practically, this is the time personnel were on or in the vehicle or were working on the vehicle. If this time is not known, an estimate of the total time spent conducting Level II or Level III activities can be used. For ingestion scenarios, separate dose rates were derived for Level II and Level III scenarios. Appropriate rates should be used to estimate intake rates, E(50) rates, and peak kidney uranium concentrations from the actual exposure scenario information provided.

There are caveats for the use of these values. The intake and dose rates presented in the table were obtained under the assumption that protective practices were not employed and the vehicle had not been decontaminated. If respiratory protection was used, the inhalation values need to be reduced by the protection factor provided by the respiratory protection worn. If gloves were worn and hand washing occurred, the ingestion values will significantly overestimate the actual ingestion. If hand monitoring occurred and the monitoring indicated no contamination, it is recommended that the hand-to-mouth ingestion rate be reexamined and either estimated based on the minimum detection capability of the instrument used or set to zero based on professional judgment.

If multiple exposure duration times are provided, *the peak kidney uranium concentration rate cannot be simply multiplied by the total time,* because uranium clearance from the kidney will occur between the exposures. Therefore, if multiple exposures occur, the intervals of exposure and the time between the exposure intervals are necessary to estimate the peak kidney concentration. The E(50) for multiple exposure duration scenarios can be calculated by multiplying the E(50) rate by the total time, because E(50) is cumulative.

4.2 SUMMARY OF LEVEL II AND LEVEL III EXPOSURES

This section summarizes the Level II and Level III estimates of intake rates, committed dose rates, and peak kidney uranium concentrations. The reported estimates are based on Capstone data when available and estimates reported in USACHPPM (2000). Capstone data were used to update the inhalation and ingestion exposure estimates for Level II and Level III personnel.

4.2.1 Level II Estimates

Table 4.8 summarizes the estimated intake rates, committed dose rates, and peak kidney uranium concentrations per hour for Level II personnel who spent time in and around a DU-contaminated vehicle.

The table includes the Capstone inhalation and hand-to-mouth ingestion exposure estimates (procedures discussed above) and comparable values reported by USACHPPM (2000). Estimates developed by USACHPPM (2000) for personnel engaged in special work procedures (welders repairing DU-contaminated vehicles and personnel removing the DU armor package from an Abrams tank). USACHPPM (2000) provides a detailed summary of

Table 4.8. Best Estimates of Level II Exposures from Capstone Data and USACHPPM Data Assuming No Use of Personal Protection Equipment

Source of Data	Intake Route	Intake Rate (mg/h)		Committed Dose Rate (rem/h)		Peak Kidney U Concentration (µg U/g kidney-h)	
		Lower Estimate[a]	Upper Estimate[b]	Lower Estimate[a]	Upper Estimate[b]	Lower Estimate[a]	Upper Estimate[b]
General Level II Recovery Activities							
Capstone Study (Parkhurst et al. 2004)	Inhalation	0.447	14.5	1.97E-03	7.80E-02	2.89E-03	1.43E-01
	Ingestion	10.6	10.6	7.07E-04	7.07E-04	7.67E-02	7.67E-02
	Total	11.0	25.1	2.68E-03	7.87E-02	7.96E-02	2.20E-01
USACHPPM (2000)	Inhalation	7.8E-04	2.5E-02	1.0E-05	5.0E-04	3.2E-05	6.7E-04
	Ingestion	2.8E-02	5.7E-02	2.0E-06	2.0E-06	4.7E-04	9.5E-04
	Total	2.9E-02	8.2E-02	1.2E-05	5.0E-04	5.0E-04	1.6E-03
Special Procedures							
USACHPPM (2000)	Welding (No Protection)	1.5E-03	0.14	3.0E-05	3.0E-03	8.0E-05	5.5E-03
	DU Armor Package Removal	2.4E-03	0.56	5.0E-05	1.0E-02	6.6E-05	1.4E-03

(a) Represents estimate of Level II exposure per hour of recovery operations with time spent inside and outside the vehicle.
(b) Represents estimate of Level II exposure assuming personnel are inside the vehicle 100% of the time.

the methods used to estimate exposures associated with special work procedures. The values in Table 4.8 are applicable to personnel who work in contaminated vehicles for extended periods due to their job functions, including those involved in battle damage and assessment, explosive ordnance disposal, maintenance, and supply and transport. Upper bound rate estimates are listed for time spent inside the vehicle, and lower bound rate estimates are listed for time spent on and around the vehicle. *The estimated values assume no use of personal protective equipment such as respirators and no vehicle cleanup or decontamination before entry occurs.*

The large difference between the lower estimate of exposure and the upper estimate for these Level II exposures results from the way the estimates were calculated and what the numbers represent. As previously discussed, the Capstone inhalation lower estimate was based on breathing zone samplers worn by personnel conducting sample recovery operations. The data from these breathing zone samplers represent a mean exposure for an entire Level II-like process—part of the exposure was received while approaching the vehicle, part while working inside the vehicle, and part when exiting the vehicle. The hourly rate represents the exposure rate per hour of Level II-like operations, which includes the time spent both inside and outside the vehicle. This estimate is based on recovery operations conducted during all three phases.

The upper estimate represents the exposure rate for only the time spent inside the vehicle. These data were obtained from instrumentation that was positioned inside an Abrams tank and was operating during sample recovery operations. Because resuspension is based in part on the amount of surface contamination inside the vehicle, the resuspension rates in a Bradley vehicle should be less, whereas the resuspension rates for an unventilated Abrams tank with DU armor perforated by a DU munition should be somewhat higher. The rates for any vehicle with its ventilation systems operating at the time of perforation or turned on shortly after perforation should be significantly less.

The estimated DU intake rates for Level II personnel, based on the Capstone Study, are low but are in the range where protective measures may be warranted for personnel who routinely work inside vehicles that contain DU residues. The total (inhalation and secondary ingestion) radiological dose rate for Level II personnel ranged from a lower estimate of approximately 3×10^{-3} rem/h to an upper estimate of approximately 8×10^2 rem/h for personnel working without protective equipment (primarily respiratory protection) or on vehicles that had no prior cleanup or decontamination.

Estimates of peak kidney uranium concentration rates from these exposures ranged from a lower estimate of approximately 8×10^2 μg U/g kidney-h to an upper estimate of approximately 2×10^{-1} μg U/g kidney-h.

4.2.2 USACHPPM Level II Results of DU Munitions Cleanup

Personnel involved in cleanup activities associated with the Camp Doha explosion and fires that occurred after ODS are also a part of Level II exposure. Camp Doha was a major U.S. Army depot that had DU munitions both in storage areas and uploaded in Abrams tanks. On July 11, 1991, a fire engulfed Abrams tanks uploaded with DU munitions in the Camp Doha motor pool. The fire quickly spread to other vehicles, ammunition carriers, and ammunition storage areas in the Camp Doha motor pool. USACHPPM (2000) has a detailed account of the fire and the aftermath of the fire.

Table 4.9 summarizes the estimated radiation doses and kidney uranium concentrations to personnel involved in the Camp Doha cleanup operation as detailed in USACHPPM (2000). The values are the estimated doses for the entire cleanup operation and not the dose rates as reported above. The doses assume that no protective equipment was worn and were estimated for each group of personnel who participated in the cleanup. Groups included explosive ordnance disposal personnel, survey personnel, and cleanup personnel who were involved in physically shoveling and sweeping the debris.

As shown in the table, the radiation doses and kidney uranium concentrations vary widely. All kidney uranium concentrations would be in Renal Effects Group (REG) 0 for which no effects to the kidney would be predicted. The radiological doses would be in radiation exposure status (RES) 0 or 1A (see Chapter 5 for more details or Chapter 6 of Attachment 3 for complete details regarding RES categories of risk). Additionally, all of the doses and kidney uranium concentrations are below any of the established occupational protection standards.

Table 4.9. Estimated Radiation Doses and Kidney Uranium Concentrations for Camp Doha Cleanup Personnel

Metric	Radiation Dose (rem)			Kidney Uranium Concentration (μg U/g kidney)		
	Inhalation	Ingestion	Total	Inhalation	Ingestion	Total
Mean	1.1E-02	1.5E-03	1.3E-02	7.0E-03	3.5E-02	4.2E-02
Standard Deviation	2.0E-02	1.8E-03	2.2E-02	1.1E-02	2.3E-02	3.1E-02
Minimum	8.2E-04	2.2E-04	1.1E-03	9.3E-04	2.2E-03	3.4E-03
Maximum	6.0E-02	5.5E-03	6.6E-02	3.5E-02	6.3E-02	9.8E-02

4.2.3 USACHPPM Level III Results

Table 4.10 is a summary of the USACHPPM (2000) Level III results with the exception of the exposures for personnel inside vehicles struck by DU munitions. The Capstone Level II results were used for this estimate. The Capstone study did not yield data that could be used to update any of the other data points. USACHPPM (2000) based its exposure estimates for personnel exposed outside a vehicle on the robust data developed by Fliszar et al. (1989). USACHPPM (2000) has detailed descriptions of the assumptions used for each of the values listed below. These data are presented as doses per hour of exposure. Care must be taken when using the kidney data because, unlike the model for radiation effects, the chemical effects of uranium exposure are based on the peak concentration of uranium in the kidney. For a chronic exposure or multiple exposures, the radiation doses can be added together to estimate the overall effect. This can not be done for chemical effects, because uranium is excreted from the kidney over time.

Table 4.10. Level III Exposure Estimates from USACHPPM (2000) (except where otherwise indicated)

Source	Exposure Pathway	Intake (mg/h)		Dose, E(50) (rem/h)		Kidney U Concentration (μgU/g kidney-h)	
		Lower Estimate	Upper Estimate	Lower Estimate	Upper Estimate	Lower Estimate	Upper Estimate
Downwind of a Burning, Uploaded Abrams Tank	Inhalation	3.9E-06	2.8E-03	6.0E-08	4.0E-05	1.0E-07	7.0E-05
Entry of Burned Uploaded Abrams Tank	Inhalation	5.7E-03	2.5E-02	1.0E-04	4.0E-04	1.5E-05	6.7E-04
Entry of Vehicles Perforated by DU Munitions[a]	Inhalation	0.447	14.5	1.97E-03	7.80E-02	2.89E-03	1.43E-01
	Secondary Ingestion	1.78	1.78	1.20E-04	1.20E-04	1.30E-02	1.30E-02
	Total	2.23	16.3	2.09E-3	7.81E-2	1.59E-02	1.56E-01
Downwind of a Vehicle Perforated by DU	Inhalation	6.30E-05	4.40E-02	1.00E-06	7.00E-05	1.00E-06	2.00E-04

(a) From Capstone data

Table 4.11 summarizes the Level III exposures to personnel at Camp Doha who were not involved in cleanup efforts. The data presented are the total doses for the entire event and are not dose rates. Clearly, the kidney uranium concentrations and radiation doses estimated are well below the chemical toxicity guideline and U.S. NRC occupational radiation dose limits.

Table 4.11 Exposure Estimates of Camp Doha Personnel Not Involved in Cleanup and Personnel Downwind of the Fire

Exposure Type	Radiation Dose (rem)		Kidney U Concentration (µg U/g Kidney)	
	Minimum	Maximum	Minimum	Maximum
Camp Personnel	1.9E-08	6.2E-08	1.8E-09	5.9E-09
Maximum 1 km Downwind	--	3.0E-06	--	2.8E-07

4.3 COMPARISON WITH OTHER ESTIMATES

Table 4.12 compares the Level II results from The Royal Society (2001, 2002), USACHPPM (2000), and the Capstone study (Szrom et al. 2004). There was reasonable agreement for the estimates of peak kidney uranium concentration rates. The Capstone lower and upper estimates fell within The Royal Society's (2001, 2002) central and worst-case estimates. The Royal Society's worst-case kidney uranium concentration was a factor of four higher than the Capstone upper estimate. The Royal Society's central estimate for radiation dose was within the range of the Capstone lower and upper estimates. The Royal Society's worst-case estimate was five times higher than the upper estimate made from Capstone data. Given the degree of conservatism in The Royal Society's worst-case estimates, it was expected that the values would be significantly higher.

Table 4.12. Comparison of The Royal Society (2001, 2002), USACHPPM (2000), and Capstone Study (Szrom et al. 2004) Estimated Radiation Dose Rates and Peak Kidney Uranium Concentration Rates for Level II Personnel

Parameter	The Royal Society		USACHPPM 2000		Capstone Study	
	Central Estimate	Worst-Case	Lower Bound	Upper Bound	Lower Bound	Upper Bound
E(50) Rate(rem/h)						
Inhalation	5.0E-03	4.4E-01	1.0E-05	5.0E-04	1.97E-03	7.80E-02
Ingestion	5.0E-06	3.0E-04	1.0E-06	2.0E-06	7.07E-04	7.07E-04
Total (rem/h)	5.0E-03	4.4E-01	1.1E-05	5.0E-04	2.68E-03	7.87E-02
Peak Kidney U Concentration Rate (µgU/g kidney-h)						
Inhalation	5.0E-03	9.6E-01	3.2E-05	6.7E-04	2.89E-03	1.43E-01
Ingestion	3.0E-04	3.0E-02	4.7E-04	9.5E-04	7.67E-02	7.67E-02
Total (µg U/g kidney-h)	5.3E-03	9.9E-01	5.0E-04	1.6E-03	7.96E-02	2.20E-01

The agreement between Szrom et al. (2004) and The Royal Society (2001, 2002) central estimates yields increased confidence in the results presented in Szrom et al. (2004) because each study used different approaches to calculating results. The Royal Society's approach was based on theoretical assumptions concerning the percentages of the penetrator that was aerosolized, particle characteristics, and the maximum aerosol that could be internalized. Szrom et al. (2004) used experimental data from the Capstone tests (Parkhurst et al. 2004) to estimate human health risks to Level II and Level III personnel.

A comparison of the Level III exposures estimated by USACHPPM (2000) and The Royal Society is complex because of the different assumptions used. The approach taken by The Royal Society to estimate exposures for Level III personnel entering damaged or burned vehicles was to use Level II estimates that were modified for the amount of time spent inside each vehicle, whereas the USACHPPM estimates used data from Fliszar, et al. (1989).

4.4 REFERENCES

James, AC. 2002. *User Manual and Technical Basis for IMBA-URAN Cameco Edition (Version 3.0)*. ACJ & Associates, Inc., Richland, Washington.

Brouwer DH, R Kroese, and JJ Van Hemmen. 1999. "Transfer of Contaminants from Surface to Hands: Experimental Assessment of Linearity of the Exposure Process, Adherence to Skin and Area Exposed During Fixed Pressure and Repeated Contact with Surfaces Contaminated with a Powder." *Applied Occupational and Environmental Hygiene*, Volume 14: 231-239.

Davies JE, ER Stevens, and DC Staiff. 1983. "Potential Exposure of Apple Thinners to Azinphosmethyl and Comparison of Two Methods for Assessment of Hand Exposure." *Bull Environ Contam Toxicol*, 31: 631-638.

Fenske RA, SG Birnbaum, MM Methner, and R Soto. 1989. "Methods for Assessing Fieldwiorker Hand Exposure to Pesticides During Peach Harvesting." *Bull Environ Contam Toxicol*, 43: 805-813.

Fliszar RW, EF Wilsey, and EW Bloore. 1989. *Radiological Contamination from Impacted Abrams Heavy Armor*. Technical Report BRL-TR-3068, Aberdeen Proving Ground, Ballistic Research Laboratory, Aberdeen, Maryland.

Office of the Special Assistant to the Deputy Secretary of Defense for Gulf War Illnesses (OSAGWI). 1998. *Exposure Investigation Report, Depleted Uranium in the Gulf*. Online report available at URL: www.gulflink.osd.mil in the Environmental Exposure Reports Section.

Office of the Special Assistant to the Deputy Secretary of Defense for Gulf War Illnesses (OSAGWI). 2000. *Exposure Investigation Report, Depleted Uranium in the Gulf (II).* Online report available at URL: www.gulflink.osd.mil in the Environmental Exposure Reports Section.

Parkhurst MA, F Szrom, RA Guilmette, TD Holmes, YS Cheng, JL Kenoyer, JW Collins, TE Sanderson, RW Fliszar, K Gold, JC Beckman, and JA Long. 2004. *Capstone Depleted Uranium Aerosols: Generation and Characterization, Volumes 1 and 2.* PNNL-14168, Prepared for the U.S. Army by Pacific Northwest National Laboratory, Richland, Washington.

Pierce JT, M Cathy, and LW Weber. 1998. "Dermal Exposure and Occupational Dermatoses", Chapter 14 in *The Occupational Environment – Its Evaluation and Control,* DiNardi SR, editor, AIHA Press, Fairfax, VA.

Szrom F, EG Daxon, MA Parkhurst, GA Falo, and JW Collins. 2004. *Level II and Level III Inhalation and Ingestion Dose Calculations.* PNWD-3480, prepared for the US Army by Battelle under the Chemical and Biological Defense Information Analysis Center Task 241, DO 0189, Aberdeen, Maryland.

The Royal Society. 2001. *The Health Hazards of Depleted Uranium Munitions Part I.* Policy Document 6/01, London, United Kingdom. Online report available at URL: www.royalsoc.ac.uk in the Science Policy Section.

The Royal Society. 2002. *The Health Hazards of Depleted Uranium Munitions Part II,* Policy Document 5/02, London, United Kingdom. Online report available at URL: www.royalsoc.ac.uk in the Science Policy Section.

US Army Center for Health Promotion and Preventive Medicine (USACHPPM). 2000. *Depleted Uranium—Human Exposure Assessment and Health Risk Characterization in Support of the Environmental Exposure Report "Depleted Uranium in the Gulf" of the Office of the Special Assistant to the Secretary of Defense for Gulf War Illnesses, Medical Readiness and Military Deployments (OSAGWI), OSAGWI Levels I, II and III Scenarios, 15 September 2000.* Health Risk Assessment Consultation No. 26-MF-7555-00D, Aberdeen Proving Ground, Maryland. Online report available at URL: www.gulflink.osd.mil in the Environmental Exposure Reports Section.

5.0 Depleted Uranium Chemical and Radiation Risk in the Military Context

The content of this chapter is particularly for military planners, policy makers, and end users who are in charge of managing the risks to personnel engaged in military operations. Further explanations of the military radiation exposure control philosophy and possible military risk management of exposure to depleted uranium (DU) aerosols are presented in the Human Health Risk Assessment (HHRA) (Attachment 3, Chapter 7).

Federal and U.S. DoD regulations require operational commanders, at all levels, to include long-term health effects in their combat operational planning and execution to ensure that the total risk to personnel is considered. This planning and execution must include consideration of risks that might be encountered before, during, and after combat. The total risk to an individual includes those from operations, battle and non-battle injuries, those associated with protective procedures if implemented, and DU aerosols and other toxicants. The DoD has adopted a detailed risk management system that is documented in U.S. Army Field Manuals (FM) (FM 3-100.12 [2001]; FM 3-101.12 [2001]; FM 100-14 [1998]). This system relies on the commander's experience to judge the severity and the probability of a risk occurring to determine the overall risk posed by the part of the hazard that cannot be mitigated.

Occupational standards developed to protect workers from exposure to potentially hazardous materials are often conservative and are based on projected chronic exposures (i.e., 40 h/week, 50 weeks/year over an employment lifetime). Sub-clinical toxicological effects serve as the basis for many toxicity standards. The Linear, No-Threshold approach to potential long-term radiological effects serves as the basis for setting some radiological limits. In a peacetime environment, the radiation and chemical safety standards can be used in a comparative way to judge the hazard associated with chemical and radiation exposure. The same safety standards cannot be directly applied to military operational environments because the occupational risks in this environment can be radically different from the occupational environment for which these standards were developed. The key differences are briefly described below.

- The level of risk in a military operation ranges from risks that are comparable to non-combat occupational environments to risks that include injury or death.

- The application of a protective practice may incur risks that are greater than those averted. In a non-combat occupational environment, the risk associated with the use of a protective procedure is usually a matter of cost and convenience. In a military environment, this risk may be death if the measure intended to protect personnel from one hazard increases the probability of occurrence of another, more immediate hazard.

- In the military environment, the "employer" does not entirely control the risk. The adversary strongly influences when, where, how, how much, and what types of weapons, including radiological or chemical toxins, will be employed.

The commander of military operations needs a system to balance immediate risk (morbidity and mortality from combat, operational accidents, and diseases) with risks from chemical or radiological exposures that may not manifest health effects until months or even years after the combat operation.

5.1 SYSTEM FOR PLACING DU RADIOLOGICAL AND CHEMICAL RISKS INTO OPERATIONAL CONTEXT

The North Atlantic Treaty Organization (NATO) and DoD addressed the problem of radiation risk by developing guidance for commanders that establishes upper bounds for radiation exposure for a particular operation (NATO Standard Agreement 2083 [1996]; U.S. Army Joint Publication [JP 3-11] 2000). The commander's risk assessment task now focuses on determining whether exposure limits will be exceeded. Risk mitigation measures are taken to maintain exposures below the risk level set for a given military operation.

Similar NATO and DoD guidance is not available for chemical toxicity. To address this need for DU aerosols, the HHRA (Attachment 3, Chapter 6) developed a risk model to predict health effects of uranium on the kidney.

5.1.1 Radiological Health Risk

DoD and NATO have developed the policy and doctrine that provides field commanders with radiation exposure guidance for use in military operations ranging from peacetime garrison operations to global conflagration involving the exchange of nuclear weapons. The policy establishes a Radiation Exposure Status (RES) system that allows commanders to select the maximum radiation exposure allowed based upon the risks of the mission, the importance of the mission, and prior radiation exposure (U.S. Army Field Manual [FM 3-11] 2003). The RES categories are based either on "peacetime" standards/guidance (RES 0 through RES-1D) or NATO standards for operations ranging from high-intensity combat (Operation Desert Storm [ODS], Operation Enduring Freedom, and Operation Iraqi Freedom) to global conflagration (RES-1E through >RES-2). The RES categories, the associated doses, the source of the RES category, and the increased cancer risk represented by the RES category are presented in Table 5.1.

RES guidance states that "field units" involved in stability and support operations (i.e., operations that have operational risks ranging from those that

Table 5.1. Comparison of RES Categories with "Peacetime" Standards or Guidelines

RES Level	Max Dose[a] (rad or rem)	"Peacetime" Equivalent	Reference	Cancer Risk[b]
RES-0	0.05	One-half of the annual limit for the public (0.1 rem/yr, 10 CFR 20)	JP 3-11 (2000)	2.0E-05
RES-1A	0.5	Annual dose limit for a member of the public when prior authorization is obtained	10 CFR 20	2.0E-04
RES-1B	5	Annual dose limit for occupational workers	10 CFR 20	2.0E-03
		U.S. Environmental Protection Agency (EPA) Protective Action Guide (PAG) level for all workers in a radiological emergency	EPA 1992	
RES 1C	10	Annual dose limit for occupational workers with a planned special exposure	10 CFR 20	4.0E-3
		EPA PAG level for protecting valuable property during an emergency	EPA 1992	
RES-1D	25	EPA PAG level for protecting large populations or saving lives	EPA 1992	2.0E-02
RES-1E	75	No equivalent; EPA PAG for exposures greater than 25 rem for protecting large populations or saving lives on a voluntary basis – NATO	EPA 1992 / JP 3-11 (2000)	6.0E-02
RES-2	125	No equivalent – NATO	JP 3-11 (2000)	1.0E-01
RES-3	>125	No equivalent – NATO	JP 3-11 (2000)	>1.0E-01

(a) RES categories are defined in terms of rad (cGy). For gamma exposures in which the quality factor is 1, the dose in rad (cGy) is equal to the effective dose in rem (0.01Sv) for whole-body exposure.
(b) Cancer risk is calculated using the methodology outlined in HHRA (Attachment 3). (Note, a dose and dose rate reduction factor of two was applied to RES 1C and below.)

would exist under peacetime conditions to risks that would be encountered in situations just below high-intensity combat) cannot exceed RES-1E. Units involved in high-intensity combat, such as ODS or Operation Iraqi Freedom, may exceed RES-1E but should not exceed RES 2. Each of these RES levels has protective actions that are required, and commanders are provided guidance concerning when each level is appropriate. For operations in which operational risks are no higher than peacetime garrison levels, a low RES level is appropriate. As the risk of the operation increases, a higher RES level is acceptable. This system allows a commander to balance individual risks to minimize the overall risk (operational and radiation) to personnel. The HHRA (Attachment C, Chapter 7) has an extensive discussion of RES levels.

Use of the RES system for addressing radiological risks from exposure to DU aerosols from vehicle perforations is a reasonable, if conservative, approach for relating DU risks to other operations. The RES categories were developed specifically for whole body exposure to penetrating radiation (i.e., neutron and gamma-ray exposures) and are based on the ability of penetrating radiation to depress immune system response, increase susceptibility to other injuries and, at higher doses, damage other organs systems in addition to the immune system.

As documented by Mettler (1999) in a study sponsored by the Institute of Medicine, each of the RES levels has associated training, risk mitigation, individual dose recording, and occupational follow-up requirements. These requirements were developed specifically for whole body exposure by penetrating radiation (neutron and gamma-ray exposures) and did not include the effect of internal emitters because external dose was projected to be the primary hazard and the internal dose was negligible. Modifications should be made to the RES levels to make them more relevant for the type of non-uniform internal exposure generated by DU intakes. Two specific changes that should be considered are identified and discussed below.

- *Changes in the medical outcomes associated with each of the RES categories.* The Armed Forces Radiobiology Research Institute (AFRRI) (2003) provides descriptions of the medical outcomes for each of the RES categories. These outcomes are based upon the assumption that the dose was delivered to the whole body by either gamma or neutron radiation. AFRRI (2003) predicts increased susceptibility to other injuries for RES categories 1C and above. Gamma or neutron exposures in excess of RES 1C have the potential to compromise the immune system (primarily the red bone marrow) and increase susceptibility to disease and infection. DU exposures that result in radiation risks in excess of the risks associated with RES 1C do not induce similar damage to the immune system because the dose to the bone marrow is so low. The highest bone marrow dose estimated in the HHRA was less than 0.5 rem, which is well below the 10-rem level for RES 1C.

- *Changes in the recommendations for evacuation from theater.* Current RES medical guidance (AFRRI 2003) recommends evacuation of personnel in RES 1D or greater. Once again, this is due to the effect of penetrating radiation on the immune system and is not required for internalized depleted uranium.

The remainder of the requirements for documentation, dose optimization, training, and medical follow-up are required in accordance with the guidance provided for each RES category (JP 3-11 2000; AFRRI 2003).

5.1.2 Chemical Health Risk

A military system comparable to the RES system is not available for evaluating chemical toxicity. The primary concern for uranium chemical toxicity is adverse effects to the kidney. Current occupational guidelines for chemical toxicity are not based upon onset of adverse health effects. Rather, the *de facto* occupational guideline of 3 µg U/g kidney and discussions concerning lowering the guideline are based on concern for protecting sensitive populations as well as controlling the onset of subtle changes in kidney function that may be the first indicators that a harmful kidney uranium concentration is being approached.

In the combat context, "harmful" refers to a kidney uranium concentration that would cause acute illness, permanent kidney damage, or require treatment to prevent mortality or morbidity. A risk model was developed in the HHRA (Section 2.3.1 and Attachment 3) to relate kidney uranium concentrations to potential adverse renal effects. This risk model was based upon data from human exposures to uranium. For each exposure, the concentration of uranium in the kidney was correlated with the severity of renal effects noted. This risk model established Renal Effects Groups (REGs); each REG correlates a range of kidney uranium concentrations with predicted renal effects. The range of kidney uranium concentration for each REG as well as the effects and predicted outcomes are as follows:

- REG 0 is composed of kidney uranium concentrations ≤2.2 µg U/g kidney in which no detectable health effects are predicted.

- REG 1 is composed of concentrations >2.2 and ≤6.4 µg U/g kidney. Persons receiving a dose within this range are not likely to become ill. At worst, this exposure may cause transient changes in renal function.

- REG 2 is composed of concentrations >6.4 and ≤18 µg U/g kidney. Persons receiving a dose within this range may become ill. Protracted (but not permanent) changes to renal function may occur.

- REG 3 applies to concentrations >18 µg U/g kidney. Depending on the dose, severe clinical symptoms of renal dysfunction are possible.

REG 0 requires no action because neither transient nor permanent kidney effects are predicted. Similarly, no transient or permanent harm is anticipated for REG 1; however, the potential exists that the *de facto*, peacetime occupational guideline of 3 µg U/g kidney may be exceeded. This may necessitate mitigation actions to reduce exposure if the operational risks are similar to peacetime risks, and according to DoD policy, occupational follow-up would be required. Personnel in REG 2 may become ill and may have protracted indicators of kidney dysfunction. Follow-up is required and some degradation in performance may be experienced. As with REG 1, occupational follow-up is required. Personnel in REG 3 may require treatment and may have protracted kidney effects, most likely reversible. As with REG 1 and 2, occupational follow-up would be required.

5.2 OPERATIONAL IMPACT OF THE ESTIMATED EXPOSURES

The DU risks estimated in the HHRA provide information that can be used by a commander and medical personnel for pre-operational planning, conduct of operations, and post-operation follow-up. The RES categories and Renal Effects Groups are summarized in Table 5.2 for the five exposure scenarios for OSAGWI Level I exposures to crewmembers in a vehicle at the time of perforation and first responders entering shortly after perforation. Crewmember risks are divided into two major scenario categories—Most Likely and Upper Bound—based primarily on crew stay-times in the vehicles. Discussions with crewmembers indicated that during ODS, the crewmembers in perforated vehicles exited quickly, especially in Abrams tanks (Deployment Health Support Directorate 2004). Risks were calculated for most likely exposure times between 1 and 5 min. Risks for longer stay-times were also evaluated. Caveats that apply to results presented in Table 5.2 are discussed in the text that follows.

These results represent conservative, upper end exposures that probably overestimate effects for Level I exposures. The significant reduction in

Table 5.2. Summary of Radiological and Chemical Health Risk Categories

Scenario	Abrams Tank: Conventional Armor, No Ventilation		Abrams Tank: DU Armor, No Ventilation		Abrams Tank: DU Armor, EC/NBC Operating		Bradley Vehicle: Conventional Armor, No Ventilation	
	RES	REG	RES	REG	RES	REG	RES	REG
Crewmembers								
Most Likely								
A-Crew, exit in 1 min	1B**	1	1B**	0	1A	0	1B	0
B-Crew, exit in 5 min	1B**	1/2	1C	1	1B	0	1B**	1
Upper Bound								
C-Crew, exit in 1 h	1C	2	1C/1D	1*	1B	0	1B**	1*
D-Crew, exit in 2 h	1C	2	1C/1D	1*	1B	0	1B**	1*
First Responders								
E-Entry at 5 min post shot, exit 10 min later	1B	0	1B**	0	1B	0	1B	0

1*- Renal Effects Group 1 that exceeded the *de facto* occupational guideline of 3 µg U/g kidney.
1A*- RES-1A that exceeded the risk associated with the 0.1 rem/yr limit for the general public.
1B** - RES 1B that exceeded the risk associated with ICRP (2 rem/yr occupational level, averaged over 5 years).

effects from use of an Abrams tank environmental control/nuclear, biological, chemical (EC/NBC) ventilation system is illustrated in the results shown in the Table 5.2 column titled "Abrams Tank (Through DU Armor, EC/NBC ventilation system on)." Similar reduced effects, though probably less dramatic, would be expected for a Bradley vehicle with an operating ventilation system at the time of perforation. The impact of a Halon fire-suppression system on exposures is not considered in any of the risk estimates, and most likely, activation of the fire-suppression system would serve to reduce the availability of aerosols for inhalation.

The estimates discussed above are useful for risk assessment purposes and can be used to assign temporary RES status for personnel. However, they should not be used to estimate individual doses primarily because individual doses are affected by many variables (such as breathing rate, exposure duration, and particle solubility). Additionally, the potential exists for other sources of internalized DU, such as wound contamination or the presence of embedded fragments. Instead, biomonitoring, usually by urine bioassay, should be used to assign individual dose. The delay between exposure and biomonitoring should be managed to ensure that significant intakes of DU can be detected. "Significant" in this case refers to intakes that would approach or potentially exceed occupational radiation and/or chemical standards or may approach the level at which peacetime protection practices require biomonitoring. USAMEDCOM (2003) contains guidance for the allowable time delay between the exposure and bioassay. The information presented in this assessment can be used as guidance for those military personnel populations that may meet the "significant" criteria.

Several conclusions can be drawn from the information in Table 5.2 concerning Level I exposures. First, the radiation risks and the chemical risks are low enough that, for combat operations, crew-response activities should not be altered because of the presence of DU in or on a vehicle. Rather, only those risk mitigation procedures that do not compromise an individual's ability for self-protection should be undertaken. An effective action that can be taken to protect crewmembers is turning on ventilation systems. The HHRA showed that it was unlikely that any crewmember or first responder exceeded the risks associated with U.S. peacetime exposure limits for saving life (RES 1D). It is probable that they did not exceed the risks associated with the U.S. annual occupational exposure limit of 5 rem/yr (RES 1B). While crewmembers may have exceeded the *de facto* kidney occupational guideline of 3 µg U/g kidney, it is unlikely that a crewmember or a first responder would receive a kidney uranium concentration that would result in severe adverse renal effects with the possible exception of those crew in an Abrams tank (perforated through conventional armor) for 5 min or more. From an operational perspective, a personnel in RES category of at least RES 1C should be expected if the potential exists that DU munitions will be used in combat.

Second, as with many of the hazards on the battlefield, the risk levels for crewmembers are high enough that hazard awareness training should be conducted and appropriate post-operation medical follow-up (biomonitoring) should be conducted to assess individual exposures.

First responder risks are below those associated with the peacetime annual exposure limits. Personnel within this category are not expected to receive kidney uranium concentrations in excess of the 3 µg U/g kidney *de facto* occupational guideline. Furthermore, the kidney uranium concentrations are within REG 0, no detectable health effects predicted. The first responder risk estimates are probably overestimates because the effect of opening the hatches for entry and exit is not included in the estimates. Like the crewmembers, the risks are sufficient to warrant post-operation biomonitoring to document the exposures received as required by peacetime occupational protection practices.

The risks to Level II and Level III personnel cannot be as easily quantified because they are dependent primarily on the amount of time personnel spent inside vehicles with DU residues, the level of contamination (i.e., whether there has been any cleanup/decontamination), the level of respiratory protection, knowledge of time spent inside the vehicle, and whether good hygiene is observed to reduce the amount of hand-to-mouth transfer of DU. Based on the results shown in Table 4.7, the risks to Level II/III personnel are low and can be readily mitigated by cleaning dust and residues from interior surfaces, limiting the amount of time spent inside a vehicle, or wearing standard respiratory protection and practicing good hygiene (using gloves if practicable and washing hands). Table 4.6 provides the recommended methodology for estimating Level II and Level III exposures based upon the information known at the time.

5.3 REFERENCES

10 CFR 20. Code of Federal Regulations, Title 10, *Energy*, Part 20, "Standards for Protection Against Radiation."

Armed Forces Radiobiology Research Institute (AFRRI). 2003. *Medical Management of Radiological Casualties*. Bethesda, Maryland.

Deployment Health Support Directorate (formerly OSAGWI), US Department of Defense. 2004. Letter from COL D. Sulka, Director, Force Health Protection, to LTC MA Melanson, US Army Center for Health Promotion and Preventive Medicine, April 22, 2004.

Mettler FA Jr, Committee Chairman. 1999. *Potential Radiation Exposure in Military Operations: Protecting the Soldier Before, During, and After*. Committee on Battlefield Radiation Exposure Criteria, Institute of Medicine, National Academy Press, Washington, DC.

North Atlantic Treaty Organization (NATO). 1996. *Commander's Guide on Nuclear Radiation Exposure of Groups. North Atlantic Treaty Organization (NATO) Standardization Agreements*. STANAG-2083, Edition 5, (originally issued September 19, 1986 with latest amendment issued June 26, 1994), Brussels, Belgium.

US Army Field Manual (FM) 3-100.12. 2001. "Risk Management: Multiservice, Tactics, Techniques, and Procedures," February 2001.

US Army Field Manual (FM) 3-101.12. 2001. "Multi-Service Tactics, Techniques and Procedures Manual, Risk Management," FM3-101-12, MCRP 5-12.1C, NTTP 5-03.5, AFTTP(I) 3-2.34, February 2001.

US Army Field Manual (FM) 100-14. 1998. "Risk Management," April 1998.

US Army Field Manual (FM) 3-11. 2003. "Multiservice Tactics, Techniques, and Procedures for Nuclear, Biological and Chemical Defense Operations." March 2003.

US Army Medical Command (USMEDCOM). 2004. OTSG/MEDCOM Policy Memo 03-007, 13 January 2004, "Medical Management of Army Personnel

Exposed to Depleted Uranium (DU)." Office of the US Army Surgeon General, US Army Medical Command Policy Memorandum, Fort Sam Houston, Texas.

US Environmental Protection Agency (EPA). 1992. *Manual of Protective Action Guides and Protective Actions for Nuclear Incidents*. EPA 400-R-92-001, Office of Radiation Programs, Washington, DC.

US Joint Chiefs of Staff, Joint Publication 3-11, "Joint Doctrine for Operations in Nuclear, Biological, and Chemical (NBC) Environments," JP 3-11, 11 July 2000.

6.0 Putting the Results into Perspective

The Capstone test series (Parkhurst et al. 2004, Attachments 1 and 2) was designed to support a human health risk assessment by characterizing depleted uranium (DU) aerosols resulting from perforating armored vehicles. The DU Aerosol Human Health Risk Assessment (HHRA) (Guilmette et al. 2004, Attachment 3) was designed to provide:

- Veterans of the 1991 Gulf War and other personnel who were exposed to DU aerosols (from combat, personnel and equipment recovery, or cleanup actions) with an understanding of the general magnitude of exposure to and the health risks from DU aerosols.

- Military planners, medical personnel, and field commanders with information to incorporate DU into risk assessment and risk mitigation decisions for all operational environments.

- Estimates of risk that can be used to modify U.S. DoD policies, doctrine and training for exposure mitigation, exposure monitoring and documentation, and hazard awareness training.

Depleted uranium and the DU oxide powder formed from impacting armor are not innocuous nor are they a "deadly poison." The health risks of DU oxides are comparable to many other common materials found on the battlefield and inside vehicles struck by any munitions. Because DU is a heavy metal and weakly radioactive (although less radioactive than natural uranium), it has received considerable attention, much of which has focused on potential adverse health effects. As with most materials, it is the quantity taken into the body and the route of entry that determine whether it will cause an adverse health effect. Natural uranium is ubiquitous in earth's soils and is found naturally in our food and water supplies. No adverse health effects have been traced to these natural levels of uranium.

Uranium has been widely studied, and it is known that at sufficiently high doses, it can cause kidney damage. There is no compelling evidence that DU causes cancer. Epidemiological studies of humans exposed to natural uranium, which is more radioactive than DU, indicate that cancer may result from exposure to radioactive decay products of uranium such as radium and radon, but not from exposure to the uranium itself or with its immediate progeny. Radium and radon and their progeny are not found in significant amounts in DU residues. (Details about the characteristics of DU are in Chapter 2.) Nevertheless, because DU is radioactive, the development of cancer is theoretically possible. The most relevant medical study to date of health effects from DU exposure is the study conducted by Department of Veterans Affairs. In the medical surveillance follow-up of personnel with embedded DU fragments (who would also have inhaled DU aerosols), thus far there is no clinical evidence of adverse effects from their uranium exposure (McDiarmid et al. 2004).

6.1 LEVEL I EXPOSURES

The objective of the Capstone DU Aerosol Characterization and Risk Assessment Program was to characterize the DU internalized by the most highly exposed personnel—the Level I personnel. Level I personnel are composed of two subpopulations: those who were in, on, or near an armored vehicle when the vehicle was perforated by a large-caliber DU munition, and first responders who entered the struck vehicle to render aid.

The aerosol characterization tests were designed to be as realistic as possible. Where compromises were required with field measurements, they tended to be conservative, creating conditions that would overestimate DU aerosol concentrations. For example, for logistical reasons ballistic hulls and turrets without ventilation systems were used as target vehicles in three of the four test series. Because of the lack of ventilation and the fewer surfaces available for deposition, DU aerosol concentrations were higher in the turrets in these test series than they would be in a fully operating vehicle with functional ventilation systems. In the fourth test series, the environmental control/nuclear, biological, chemical (EC/NBC) ventilation systems in an Abrams tank operated and significantly reduced DU concentrations. Again for logistical reasons, a Halon fire-suppression system, which would also reduce aerosol concentrations, was not present (Phases I through III) or did not activate (Phase IV) in the tests.

For the HHRA, predictions of health effects to Level I personnel are based on the chemical and radiological data from the Capstone study and from the current understanding of chemical and radiological doses and risks. The analysis was intended to be as realistic as possible, but some conservative assumptions were used, especially relating to radiological risks so the values calculated may overestimate the doses and risks.

6.1.1 Crewmember Exposure Incidents

A review of Operation Desert Storm (ODS) experience (Deployment Health Support Directorate 2004) shows that the Capstone experimental conditions approximated the upper bound of what actually occurred in ODS. During ODS six Abrams tanks were involved in friendly fire incidents, and in three of those six cases, a crew compartment was perforated. Most of the surviving crew exited quickly (within two minutes) or were outside at the time of the incident. The ventilation system was operating in at least one of the three cases, the Halon fire-suppression system activated in these three tanks, and hatches were opened to allow exit. This information suggests that up to 10 surviving Soldiers were exposed to DU aerosol conditions similar to, but less severe than those simulated during the Capstone Phase I field tests, in which a large-caliber DU munition perforated an Abrams tank through conventional armor.

Similarly, fifteen Bradley Fighting Vehicles (Bradley vehicles) were perforated through crew compartments by large-caliber DU munitions. Bradley crewmembers were inside 14 of the 15 perforated vehicles. Fewer details are available about crew exit times and activation of ventilation in these incidents. Some of these vehicles were hit more than once, and the Halon fire-suppression system was activated in some of these incidents.

6.1.2 Predicted DU Concentrations and Doses for Crewmembers and First Responders

The HHRA used four scenarios to model crew exposures and a fifth scenario for first responders. Scenarios A and B for personnel who exited within 1 min and 5 min, respectively, represent most likely times of exposure. Scenarios C and D apply to personnel who were exposed for up to 1 h and 2 h, respectively. These longer exposure times reflect upper bound estimates, which are listed in Chapter 3. Scenario E was developed for first responders and is based on an entrance time of 5 min after vehicle perforation and an exit up to 10 min later. The scenarios assumed that crewmembers would take no action to clear the air in the vehicle, such as activating the ventilation system or opening hatches for the duration of the stay-time.

The peak kidney uranium concentrations, radiological doses, and radiological risk of developing fatal cancer as a result of a single shot exposure for the most likely scenarios for Level I personnel are summarized in Tables 6.1 and 6.2, respectively. The risk estimates in these tables are the fatal cancer risks from DU aerosol inhalation. Comparing the 5-min and 2-h intakes in Table 3.2 shows that for most vehicle configurations, the majority (over 60%) of the intake is received during the first 5 min of exposure. Although higher intakes and associated risks occur with longer exposure durations, the effect is not linear because of the relatively rapid settling of the aerosols inside the vehicle. The rate of aerosol clearance within the vehicle is significantly increased if ventilation systems are turned on or the hatches are opened.

The Capstone data was used to predict DU peak kidney uranium concentrations, committed effective doses, and committed equivalent doses, all of

Table 6.1. Predicted Median Peak Kidney Uranium Concentrations, Doses, and Risks to Vehicle Crewmembers

Effect	Crewmembers (Most Likely Scenarios)							
	Abrams Tank: Conventional Armor, No Ventilation		Abrams Tank: DU Armor, No Ventilation		Abrams Tank: DU Armor, EC/NBC Operating		Bradley Vehicle: Conventional Armor, No Ventilation	
	1 min	5 min	1 min	5 min	1 min	5 min	1 min	5 min
Peak Kidney U Conc. (μg U/g kidney)	3.0	6.4	1.1	2.6	0.05	0.23	1.0	2.9
Committed Effective Dose E(50) (rem)	2.0	3.7	2.2	6.0	0.09	0.44	0.59	1.7
Committed Equivalent Dose H_{Lung}(50) (rem)	14	32	18	44	0.66	3.3	5.2	14
Increase in Lifetime Cancer Risk (%)	0.11	0.20	0.12	0.32	0.005	0.025	0.034	0.099

Table 6.2. Predicted Median Peak Kidney Uranium Concentrations, Doses, and Risks to Vehicle First Responders

Effect	First Responder (entry at 5 min, exit at 15 min post perforation)			
	Abrams Tank: Conventional Armor, No Ventilation	Abrams Tank: DU Armor, No Ventilation	Abrams Tank: DU Armor, EC/NBC Operating	Bradley Vehicle: Conventional Armor, No Ventilation
Peak Kidney U Conc. (μg U/g kidney)	1.5	0.67	0.14	1.4
Committed Effective Dose, E(50) (rem)	0.92	1.9	0.41	0.89
Committed Equivalent Dose, H_{Lung}(50) (rem)	8.8	14	3.1	6.7
Increase in Lifetime Cancer Risk (%)	0.050	0.10	0.023	0.052

which vary by exposure duration, vehicle configuration, and operation of ventilation systems. Each of the median results from Scenario A and E type exposures are within annual occupational exposure guidelines or limits. Some of the Scenario B results exceed the *de facto* occupational guideline of 3 µg U/g kidney (and the 5 min Abrams/non-DU armor exposure is at the upper boundary of the "not likely to become ill" category), but the concentrations are below levels of concern. The theoretical lifetime risk of fatal cancer from DU inhalation at these levels is predicted to range from 0.005 to 0.32% for crewmembers and 0.023 to 0.10% for first responders. To put these risks in perspective, the natural incidence of fatal cancers in a lifetime for males is approximately 24%. This means that for the largest dose in Table 6.1 or 6.2 (see column titled "Abrams Tank, DU Armor, No Ventilation"), the risk would increase from 24% to 24.3%. No illness from chemical toxicity is predicted, and no radiological effects are expected from any of these levels.

The DU concentration and dose levels predicted by this risk assessment suggest that most of the exposures to personnel meeting the Level I criteria are not high enough to lead to adverse health effects. In many cases the levels are below occupational dose limits and guideline values. However, the predicted intakes are high enough to warrant quantifying an individual's dose through biomonitoring. The Capstone study showed that activating the vehicle's EC/NBC or other ventilation system can significantly reduce exposures to DU aerosols. By extension, this would also reduce exposures to the other materials generated when a tank is perforated by any munition.

For cases in which the crew compartment is perforated twice, the peak kidney uranium concentrations and radiation doses can be roughly approximated by multiplying the results of a single perforation by two. The possibility of becoming ill (from transient or protracted kidney effects) increases, depending on the number of perforations and the vehicle configuration.

Care should be exercised in applying additional "safety factors" in conjunction with this assessment because of the conservative assumptions already built into the assessment. Similarly, these estimates should not be used to quantify or assign doses to individuals. Instead, biomonitoring, primarily by urine bioassay for isotopic uranium, should be used to assign individual doses. The delay between exposure and biomonitoring should be minimized to ensure that significant intakes of DU can be detected.

6.2 LEVEL II AND LEVEL III EXPOSURES

Level I exposures, by definition, occur in combat where the operational risks are extreme. In contrast, Level II and Level III exposures occur throughout the spectrum of military operations and encompass a range of operational risks that vary from the extreme risks commensurate with Level I exposures to the minimal risks associated with garrison-type operations. The level of acceptable risk from DU and the precautions taken should make sense when compared with operational risks. This report uses the analysis conducted by USACHPPM (2000) and some of the data generated during the Capstone to project to provide estimates of the Level II and Level III exposures.

The most important factor for dose to personnel who meet the criterion for either Level II or Level III exposure is the time spent inside vehicles that have not been cleaned. When operational risks allow, precautions should be taken to minimize exposure, such as wearing protective outer garments, using respiratory protection, practicing good personal hygiene (to avoid ingestion), and cleaning the vehicle.

Level II personnel have the potential for exceeding the dose levels that require biomonitoring and DU-related radiation protection training specific to their occupations. Special attention should be given to the training and the precautions recommended for explosive ordnance personnel because of the extreme risks they sometimes face when clearing a damaged vehicle. Level III personnel (as defined in this report) do not require biomonitoring but should have general training pertaining to potential DU exposures and how to reduce one's exposure and associated risks.

6.3 SUMMARY

This analysis has categorized exposures to DU aerosols similar to (but typically more severe than) circumstances encountered during ODS and has quantified the chemical concentrations and radiological doses for personnel in these categories. No attempt was made to quantify the exposures and associated doses that specifically apply to any individual. More appropriate methods exist for retrospective dose assessment including the use of personal monitoring methods. Urine biomonitoring is the most easily performed, although time limits apply, particularly for quantifying small intakes.

The doses and risks to human health from inhaling DU aerosols in a perforated vehicle are relatively low when compared with many other combat risks. In addition to the possibility of combat injuries, hazardous materials including different heavy metals, chemicals, and other agents may be released onto the battlefield and into the environment.

The most important factors for reducing exposure are 1) the use of onboard vehicle ventilation or 2) exiting quickly if circumstances allow. Even though the risks from inhaled DU based on the Capstone data are low, ventilation systems operating during or turned on as soon as possible after a DU perforation can further significantly reduce the risks to the crewmembers. As seen in Table 6.1, the reduction in risk can be greater than one order of magnitude when the EC/NBC ventilation system is activated.

The Bottom Line

After more than a decade of medical surveillance of veterans from ODS who have or had DU embedded fragments, no adverse chemical or radiological health effects related to the presence of DU have been identified. In addition to their uptake of DU through fragment wounds, these veterans would also have inhaled DU oxides and may have ingested incidental quantities of DU oxides at the time they were wounded.

The levels of DU intake predicted by the HHRA are not likely to cause adverse health effects. However, DU metal fragments and oxide powder need to be respected to avoid or minimize inhalation of resuspended aerosols and ingestion of deposited material. At sufficiently high levels, DU intake can cause adverse health effects, especially to the kidney. Except for extreme exposure conditions in which treatment may be indicated, any kidney effects would be expected to be temporary.

Although chemical and radiation risks are predicted to be low, training needs to be provided to those with a potential to be exposed to help reduce risks and keep them in perspective. Counseling of affected personnel and their family members is suggested because of the perceived radiological and chemical toxicity risks associated with exposure to DU. Because of differences in individual exposure for a given crew in a perforated vehicle, DU biomonitoring is needed to estimate doses to individuals.

The results of the HHRA suggest using the following general guidance for personnel inside an armored vehicle perforated by a suspected DU munition or inside a tank perforated through DU armor:

- If a penetrator perforates the vehicle while the EC/NBC ventilation system is off, the most effective action is to turn the system or systems on.

- If conditions outside the vehicle are safe, exit the vehicle as soon as possible and avoid immediate reentry.
- If it is unsafe to exit the vehicle (i.e., a threat exists of being fired upon during a firefight), remain in the vehicle.

6.4 REFERENCES

Deployment Health Support Directorate (formerly OSAGWI), US Department of Defense. 2004. Letter from COL D. Sulka, Director, Force Health Protection, to LTC MA Melanson, US Army Center for Health Promotion and Preventive Medicine, April 22, 2004.

McDiarmid MA, S Engelhardt, M Oliver, P Gucer, PD Wilson, R Kane, M Kabat, B Kaup, L Anderson, D Hoover, L Brown, B Handwerger, R Albertini, D Jacobson-Kram, C Thorne, and K Squibb. 2004. "Health Effects of Depleted Uranium on Exposed Gulf War Veterans: A 10-Year Follow-Up." *J. Toxicol. and Environ. Health*, 67:277-296.

US Army Center for Health Promotion and Preventive Medicine (USACHPPM). 2000. *Depleted Uranium—Human Exposure Assessment and Health Risk Characterization in Support of the Environmental Exposure Report "Depleted Uranium in the Gulf" of the Office of the Special Assistant to the Secretary of Defense for Gulf War Illnesses, Medical Readiness and Military Deployments (OSAGWI), OSAGWI Levels I, II and III Scenarios, 15 September 2000*. Health Risk Assessment Consultation No. 26-MF-7555-00D, Aberdeen Proving Ground, Maryland. Online report available at URL: www.gulflink.osd.mil in the Environmental Exposure Reports Section.

Abbreviations and Acronyms

AED—aerodynamic equivalent diameter
AFRRI—Armed Forces Radiobiology Research Institute
AM—cascade impactor array monitor
AMAD—activity median aerodynamic diameter
APG—Aberdeen Proving Ground, Maryland
ATC—Aberdeen Test Center
ATSDR—Agency for Toxic Substances and Disease Registry
BFV—Bradley Fighting Vehicle (Bradley vehicle)
BHT—ballistic hull and turret
BK—Biokinetic model
Bq/μg—becquerel(s) per microgram
BZM—personal cascade impactor breathing zone monitor
CED—committed effective dose
CEDE—committed effective dose equivalent
cGy—centiGray (equivalent to rad)
CI—cascade impactor
cm—centimeter(s)
CY—cyclone
d—day
DCF—dose conversion factor

DoD—U.S. Department of Defense
DOE—U.S. Department of Energy
DQO—data quality objective
DU—depleted uranium
DUR-IPT—Depleted Uranium Research—Integrated Process Team
E(50)—50-yr committed effective dose
EC/NBC—environmental control/nuclear, biological, chemical ventilation system
EPA—U.S. Environmental Protection Agency
ET—extrathoracic
FM—U.S. Army Field Manual
FS—filter cassette sample (or sampler)
g—gram
GI—gastrointestinal
h—hour(s)
H_T—dose equivalent to tissue T
$H_T(50)$—50-yr committed equivalent dose to tissue
HHRA—human health risk assessment (usually referring to Attachment 3 of this Summary Report)
HRTM—Human Respiratory Tract Model

ICRP—International Commission on Radiological Protection

IMBA-URAN—Integrated Modules for Bioassay Analysis-Uranium (computer code)

IOM—Institute of Occupational Medicine (used as an abbreviation for filter cassettes)

ISCORS—U.S. Interagency Steering Committee on Radiation Standards

JP—U.S. Army Joint Publication

L—liter(s)

LANL—Los Alamos National Laboratory

LC—large-caliber (referring to depleted uranium munition)

Lpm or L m^{-1}—liter(s) per minute

LRRI—Lovelace Respiratory Research Institute

m—meter

μCi—microcurie

μm—micrometer

mg—milligram(s)

min—minute(s)

MVF—moving filter

NAS—National Academy of Sciences

NATO—North Atlantic Treaty Organization

NBC—nuclear-biological-chemical

nCi—nanocurie

NCRP—National Council on Radiation Protection and Measurements

NRC—U.S. Nuclear Regulatory Commission

ODS—Operation Desert Storm

OEG—Operational Exposure Guidance

OSAGWI—Office of the Special Assistant for Gulf War Illnesses, Medical Readiness, and Military Deployment

P-CI—personal cascade impactor

PAG—protective action guides

PNNL—Pacific Northwest National Laboratory

PPE—personal protective equipment

REG—Renal Effects Group

rem—special unit of dose equivalent (U.S. conventional unit)

RES—radiation exposure status

sec—second(s)

SEM—scanning electron microscopy

SI—international system of units

Sv—sievert (special unit of dose equivalent, SI unit)

U—uranium

U.S.—United States

USACHPPM—U.S. Army Center for Health Promotion and Preventive Medicine

WHO—World Health Organization

yr—year

Glossary

Abrams Main Battle Tank—A full-tracked, armored, land combat vehicle with a 105-mm (M1) or 120-mm (M1A1/M1A2) gun operated by a four-man crew consisting of a commander, gunner, loader, and driver. The Abrams tank is the principal weapon of tank battalions of the U.S. Army during all types of combat operations.

aerosol—An assemblage of liquid or solid particles suspended in a gaseous medium long enough to be observed and measured; generally about 0.001-µm to 100-µm AED.

ballistic hull and turret (BHT)—A production Abrams or Bradley structure without any operational components. The turret is mounted on the hull via a race ring, but no other internal or external components are present (i.e., no power train, fire control system, ventilation system, etc). A BHT may contain a gun, road wheels, and track if the specific test requires these. A BHT is typically used to reduce test cost when an operational vehicle is not required to meet test objectives.

bioassay—An analysis of body fluids, tissue, or excreta to determine the absence, the degree, or presence of specific materials. Used as an index of radioactivity in the body. Urine bioassay is commonly used to monitor uranium intake.

Bradley Fighting Vehicle—A full-tracked, medium-armored fighting vehicle that provides protected, cross-country mobility and vehicular-mounted firepower to infantry/cavalry units. The Bradley Fighting Vehicle System family consists of infantry and cavalry versions.

committed effective dose, $E(\tau)$—The sum of the products of the committed organ or tissue equivalent doses and the appropriate organ or tissue weighting factor (W_T) where τ is the integration time in years following an intake. The integration time for an adult worker is 50 yr and is represented as E(50) This term replaces committed effective dose equivalent (CEDE) as defined in ICRP Publication 26.

committed equivalent dose, $H_T(\tau)$—The time integral of the equivalent dose rate in a particular organ or tissue that will be received by an individual following an intake of radioactive material into the human body where τ is the integration time in years following an intake. The integration time for an adult worker is 50 yr and is represented as $H_T(50)$. This term replaces committed dose equivalent (CDE) as defined in ICRP Publication 26.

depleted uranium—Depleted uranium (DU) is the primary material used in the large-caliber (LC) penetrators fired at vehicle targets in this study. In 10 CFR 40.4, the Nuclear Regulatory Commission defines depleted uranium as any uranium with less that 0.711 weight percent U-235. Natural uranium has approximately 0.72% by weight U-235. The depleted uranium used by the U.S. military contains approximately 0.2% by weight U-235 and is 40% less radioactive than natural uranium.

environmental control/nuclear-biological-chemical ventilation system (EC/NBC)—A system found on the Abrams Main Battle tank that conditions air for breathing (filtering out nuclear, biological, and chemical agents) as well as personal heating and cooling as required, while crew members are wearing protective suits and masks. The EC/NBC system on the Abrams tank also provides positive air pressure within the turret and driver's locations to prevent diffusion of NBC contaminants.

first responder—Personnel who enter damaged vehicles shortly after DU perforation to evacuate personnel or recover equipment.

hull—Armored structure primarily containing the power train, road wheels, and track to provide vehicle mobility.

large-caliber DU munitions—Rounds with large-caliber depleted uranium (LC-DU) penetrators that are fired from the Abrams platform (M1A1 and M1A2 series tanks). These heavy metal, long-rod penetrators use kinetic energy to penetrate a target.

penetrated—Used here to convey the piercing of the armor by the DU penetrator that may or may not enter a turret, driver, or passenger (Bradley) compartment.

perforated—Used here to convey the breach by the DU penetrator through vehicle armor into the turret, driver, or passenger (Bradley) compartment.

rem—The special unit of any quantities expressed as dose equivalent. The dose equivalent in rem (the plural is referred to as rem or rems) is equal to the absorbed dose in rads multiplied by the quality factor (1 rem=0.01 sievert).

risk—The probability that a given individual will incur a particular adverse health effect as a result of the chemical or radiation dose received.

scenario—An outline of a projected chain of events, which as it relates to this risk assessment, includes the time and duration of exposure and the breathing rate.

source term—The amount of radionuclide or chemical released from a source or site to the environment over a specific period for use in dose assessment or exposure assessment.

turret—Revolving armored structure (located on top of the tank hull) that primarily houses the main gun and fire control system.

Index

Pages containing figures are denoted by *f*. Pages containing tables are denoted by *t*.

50-yr committed effective dose. *see* E(50)
50-yr committed equivalent dose.
    see $H_T(50)$
50-yr committed equivalent lung dose.
    see $H_{Lung}(50)$

A
abbreviations, 81–82
Aberdeen Test Center, 17
Abrams tanks. *see also* vehicles
    combat perforations, ix, 23–24, 75
    definition, 83
    exposure scenarios, xix, 2–5
    health risks, 32–33, 34–36
    intakes, xx, 28, 45–49, 55–56
    test parameters, xvii–xviii, 14–15
    ventilation systems, xvii–xx, 69
acronyms, 81–82
activity median aerodynamic diameters
    (AMAD), 15
acute exposures, xxv, 54
aerosol samplers, xvii–xviii, 15, 37–38, 46
aerosols. *see also* monodisperse aerosols;
    resuspended aerosols
    concentrations, 36–38, 44, 46
    definition, 83
    dissolution rates, xix
    exposures, 4, 56
    properties, 15
Armed Forces Radiobiology Research
    Institute (AFRRI), 66–67

B
ballistic hull and turret (BHT), xvii–xviii,
    14, 83. *see also* vehicles
Battelle Memorial Institute, 4
Bayesian analysis, xvi, 6
bioassays
    definition, 83
    delay management, 69–70
    dose estimates, xxviii, 32
    isotopic uranium, 77
biomonitoring, 32, 69, 70, 74, 77–78
blood, 27
bone, 51–52
bone marrow, 20, 66–67
Bradley vehicles. *see also* vehicles
    combat perforations, ix, 24, 75
    definition, 83
    doses, 33
    intakes, 28
    test parameters, xvii–xviii, 14–15
    ventilation systems, 30, 56, 69
breathing rate, 25–26, 47–48
breathing zone monitors, 53, 56
burn tests, 2

C
Camp Doha, 43, 57 *t,* 58–59, 59 *t*
cancer. *see also* radiogenic cancer
    bone, 20
    health risks, xxii–xxiv, 19–21, 33–35,
        65, 74–77
    lung, xvii, xxii–xxiii, 19–20, 33
    lymphatic, 20
    mortality risk, xxiii–xxiv, 33–34, 34 *t,*
        35, 35 *f*

Capstone Study
  aerosol characterization, 14
  comparisons, 39 t, 39–40, 59–60
  conclusions, xxvi–xxviii, 78–80
  contamination estimates, xxiv, 49
  description, x–xi, xv–xvii, 1–5
  Level I exposures, 23, 74–77
cascade impactors (CI), 27–28, 46–49, 47 t, 48 t
central estimate, 36–37, 59
characterization, aerosols, xvii–xix, 13–17
chemical dose, xxi, 29–31
chemical risks
  health risk categories, 69 t
  models, 6
  peak kidney uranium concentrations, 29–31
  Renal Effects Groups (REG), 19, 67–68
  toxicity, xx–xxii, 17–19
cleanup, Camp Doha, 57 t
combat
  risk mitigation, 64–71
  risks, xxv–xxvi, 6, 63–64
  vehicle perforations during, 23–24, 75
committed effective dose, E(t). see E(50)
committed equivalent dose, $H_T(t)$. see $H_T(50)$
computer models. see models
concentrations, aerosol, xviii–xix, 15, 36–39, 44, 46
concentrations, kidneys
  Camp Doha cleanup, 57 t
  chemical toxicity, 29–31, 67–68
  fatal cancer risk, 76, 76 t
  inhalation exposures, 47–48
  multiple perforations, 36
  percentiles, 30–31, 31 f
  study comparisons, 36–37, 39
cotton glove contamination, xxiv–xxv, 49–53, 52 t, 54
crew compartments, 23–24, 38
crewmember health risks
  cancer, xxii–xxiv, 34 t, 34–35, 35 t, 77
  chemical, 18–21
  findings, xxvii
  peak kidney uranium concentrations, xxi, 30–31, 76 t
  radiological, 65–71
  worst case estimate, 38–40
crewmember predicted doses
  chemical, xx–xxi, 29–31
  findings, xxvii
  radiological, xxii–xxiv, 31–35
  risk estimates, 38–40, 54–59, 75–77
  upper bound calculation, 38 f
crewmembers
  exposures, xix, 23–26, 68, 75

crewmembers (cont.)
  Operation Desert Storm (ODS), ix, 1, 5, 23–24, 75
Crystal Ball (software), 50
cyclone samples, xviii, 16, 47

D
data collection
  aerosols, 5–6, 15
  cotton gloves, 15, 49–53
  gaps, x, 1, 45
  inhalation exposure, xix–xx, 25–26, 28, 46–49
  source term, xix–xx, 27–28
  uncertainty, 29, 36–38
data quality objectives, xvii, 9
decontamination, 55–57
density, depleted uranium, 13
depleted uranium, definition, 84
dissolution rates, xix, 16
dose assessments
  chemical, 6, 29–31, 36–40, 47–49, 53–60
  limits, xxii, 32–35, 58, 64–67
  occupational guidelines, xxi, 18, 30, 58, 67
  oxides, xix, 13–14, 16, 73
  radiological, xxii–xxiii, 27, 31–33, 36–40, 53–60, 75–77
  scenarios, 6, 25–26, 43–44
  study comparisons, 36–40, 59–60
  test results, xix–xxv, 27–36, 55–57
dose conversion factors (DCF), 47, 51–52, 52 t

E
E(50), definition, 83
E(50) estimates
  models, 29
  occupational limits, 31–32
  peak kidney uranium concentrations, 47–48, 54, 54 t, 58, 76 t
  sampler array data analysis, 47–48
  inside vehicles, xxii, xxii t, 32 t, 39 t
EC/NBC ventilation systems.
  see ventilation systems
EPA. see U.S. Environmental Protection Agency
epidemiology studies, 19–20
estimates
  E(50), xxii, xxii t, 31–32, 39, 47–48, 54 t, 76 t
  exposures, xix, 23–26, 58–60, 68, 75
  glove contamination, xxiv–xxv, 49–53, 52 t, 54
  intakes, xx t, 28 t, 53–55
  The Royal Society (London), 36–40, 59–60

estimates of doses
   chemical, 29–31
   inhalation, 46–49, 53 t
   Level II and III exposures, 53–60
   radiological, 31–40, 57 t, 58–59, 65–67
   stay-times, 75–76
   worst-case, 39 t
explosions and fires, 2, 24, 43–45, 57
exposure levels, xv–xvii, 2–4, 25–26, 43. see also Level I exposures; Level II exposures; Level III exposures
exposures. see also ingestion exposures; inhalation exposures; multiple exposures
   duration, xix–xx, 25–26, 43–44, 53–54
   occupational standards, 63–64
   reduction, xxiv, xxvii, 68–71, 77–80
   scenarios, xix–xx, 5, 25–26, 26 t
   whole body, 66

F
fire suppression systems, xx, 69, 74, 75
fires and explosions, 2, 24, 43–45
first responders
   chemical concentrations, xxi, 30–31, 75–76
   definition, xix, 84
   exposure scenarios, 25–26
   intakes, xx t, 28
   radiological dose estimates, xxii, xxiii, 34–35, 75–77
   risk estimates, xxiii, xxvi, 31–33, 68–69, 70, 76 t
fragments, xxviii, 3, 69, 74, 79
friendly fire incidents, 5, 23–24, 75

G
gastrointestinal tract, 27
glossary, 83–84
glove contamination, xxiv–xxv, 49–53, 52 t, 54
Gulf War of 1991. see Operation Desert Storm (ODS)

H
Halon systems. see fire suppression systems
hand contamination, 49–51, 51 f
hazard awareness training, xxvi, 70, 79
health risks
   chemical, xx xxi, 6, 17 19, 29 31, 67–69
   intakes, 79
   Level I exposures, xix–xxiv, 28–35, 75–79
   radiological, xxiii–xxiv, 19–20, 31–36

$H_{Lung}(50)$. see also inhalation exposures; lungs; respiratory tract
   crewmembers, xxiii, 33, 33 t
   ingestion exposure, 52
   peak kidney uranium concentrations, 76
   sampler array data analysis, 47–49
   worst case estimates, 39 t
$H_T(50)$, xxiii, 32–33, 52, 83–84
$H_T(t)$. see $H_T(50)$
hull, 14, 84. see also vehicles
Human Health Risk Assessment (HHRA), xix–xxiv, 28–36
   approach, 4–6
   described, 2, 8–9
   military risk assessment, 63–72
   objectives, 73
   study comparisons, 38–39

I
immune system, 66–67
impact tests, armor, 2
in vitro dissolution rates, xix, 16
ingestion exposures
   comparisons, 57–60
   duration, 54
   Level II and III estimates, xxiv–xxv, 49–57, 54 t
   triangular distributions, 50, 51 f
inhalation exposures. see also $H_{Lung}(50)$; lungs; respiratory tract
   chemical, 18
   comparisons, 57–60
   Level II and III estimates, xviii, xxiv–xxv, 53 t, 53–57, 54 t
   resuspended aerosols, 44–48
injuries and wounds. see wounds and injuries
intakes. see also ingestion exposures; inhalation exposures
   comparisons, 58
   estimates, xx, 28, 45–49, 55–59
   Monte Carlo analysis, 37
   rates, 47–48, 50, 53–57
   scenarios, xix–xx, xx t, 25–26, 28 t, 43
Integrated Modules for Bioassay Analysis (IMBA), 47
International Commission on Radiological Protection (ICRP), xix, 20, 24, 26, 36
isotopes, 12–13, 13 t

K
kidney uranium concentrations, peak
   chemical risks, 29–31, 30 t, 31 f
   crewmembers, xxi t, 36, 76, 76 t
   E(50) estimates, 47–48, 54, 54 t, 58, 76 t
   $H_{Lung}(50)$ estimates, 76
   renal effects, 29, 57, 67–68

kidneys. *see also* renal effects
   concentration rates, xxv, 37, 53–55
   dose estimates, 54 *t*, 58–59
   health risks, 18, 19 *t*, 76–77
   uranium toxicity, xxi–xxii, 17–19, 67–68

L

large-caliber munitions. *see* munitions
leukemia, 20
Level I exposures
   conclusions, 70
   definition, xv–xvi, 2–3
   description, 23–26
   military management, xxvi
   risk assessment, xix–xxiv, 29–36, 74–77
   scenarios, 68–69
   study comparisons, 36–40
Level II exposures
   definition, xv–xvi, 3, 78
   description, 43–57
   inhalation and ingestion, xxiv–xxv
   study comparisons, 60
   variables, 70–71
Level III exposures
   definition, xv–xvi, 3, 78
   description, 43–55
   estimates, 55 *t*, 58 *t*, 58–59, 59 *t*
   inhalation and ingestion, xxiv–xxv
   study comparisons, 60
   variables, 70–71
limiting exposures, 64–67, 79–80
lungs, xvii, xxii *t*, 19–20, 33. *see also* $H_{Lung}(50)$; inhalation exposures; respiratory tract
lymphatic system, 20, 27

M

MEDCOMM, xvii, 7
medical follow-up, xxvii, 32, 68–70, 74, 77–78
military operations
   exposure management, xxv–xxvi, 68–71
   Operation Desert Storm (ODS), 3, 5, 23–24, 43–45, 65–66, 75
   Operation Iraqi Freedom, 44, 65–66
military personnel
   chemical and radiation risks, 29–36, 63–71
   counseling, xxviii, 79
   dose rates, 48–49
   medical follow-up, 68–70, 74, 77–78
   inside perforated vehicles, 25, 75, 77
   physical activities, 43–45
models
   biokinetic, xvi, 19
   hand contamination dose, 49

models *(cont.)*
   inhalation exposure, xvi–xvii, 25–26
   Integrated Modules for Bioassay Analysis (IMBA), 47
   kidney uranium concentrations, 19, 67–68
   radiation doses, 27
monodisperse aerosols, 16, 47
Monte Carlo analysis, 37, 50–52, 52 *t*
mortality risks, xxiii–xxiv, 33–34, 34 *t*, 35, 35 *f*
multiple exposures, 54–55
multiple perforations, 35–36
munitions
   cleanup, 57
   definition, 84
   handling, 20–21
   penetrator properties, xxviii, 11
   performance, 13
   public debate, 1

N

National Academy of Sciences, 19–20
North Atlantic Treaty Organization (NATO), 64–65
NRC. *see* U.S. Nuclear Regulatory Commission

O

Office of the Special Assistant for Gulf War Illnesses (OSAGWI), 1–5, 44–45, 68
Operation Desert Storm (ODS)
   friendly fire incidents, 5, 23–24, 75
   Level II and III exposures, 43–45
   Radiation Exposure Status (RES), 65–66
   vehicle perforations, 3
   veterans, xxviii, 79
Operation Iraqi Freedom, 44, 65–66
oxides
   aerosols, 11
   characteristics, 13–17
   powder, 2, 73, 79
   predominant phase, xix, 16

P

P-CI. *see* cascade impactors (CI)
Pacific Northwest National Laboratory, 4
particle size, xix, 15–17, 28, 47
peak kidney uranium concentrations
   chemical risks, 29–31
   crewmembers, xxi, 76, 76 *t*
   E(50) estimates, 47–48, 54, 54 *t*, 58, 76 *t*
   $H_{Lung}(50)$ estimates, 76
   models, 19
   renal effects, 29, 57, 67–68

penetrated, definition, 84
perforations
    air samplers, 37–38
    combat, 23–24, 75
    definition, 84
    multiple penetrations, 35–36
    Operation Desert Storm (ODS), 3
    vehicles, 25, 35 f, 44–45, 66, 77
physical activities, 43–45
physiology, 26
protective equipment, lack of, 56

R
Radiation Exposure Status (RES), 65–67, 69–70
radioactive properties, 12–13
radiogenic cancer, xvii
radiological doses
    estimates, 31–40, 57 t, 58–59, 75–77
    health risks, xxii–xxiv, 31–35, 65–71, 76–77
    Level II and III exposures, 53–60
    limits, 32, 64–67
    multiple perforations, 36
radiological risks
    cancer, 19–20, 31–35
    occupational guidance, 64–67, 65 t
    risk mitigation, 70
    scenarios, 69 t
    vehicle perforations, 35 f
radium and radon, 74
Reference Man, 26
rem, definition, 84
renal effects. *see also* kidneys
    exposure limits, 18–19, 70
    peak kidney uranium concentrations, 29, 57, 67–68
Renal Effects Groups (REG), xvii, 19 t, 67–70
residues, depleted uranium, 45, 56, 71, 74
respiratory tract, 27. *see also* $H_{Lung}(50)$; inhalation exposures; lungs
resuspended aerosols, 16–17, 44–46, 48–49, 56
risk assessments
    factors effecting, 14
    Human Health Risk Assessment (HHRA), 2, 4–6, 8–9, 73
    Level I exposures, xix–xxiv, 29–36, 74–77
    military, 63–72
risk management, xxvi, 63, 70–71
risks. *see also* chemical risks; health risks; radiological risks

risks *(cont.)*
    combat, xxv–xxvi, 6, 63–64
    definition, 84
    mortality, xxiii–xxiv, 33–34, 34 t, 35, 35 f
The Royal Society (London), 36–40, 59–60

S
sample recovery, 16–17, 46–49, 53, 56
samplers, aerosol, xviii, 16, 37–38, 46
scenarios
    definition, 5, 84
    E(50), xxii–xxiii, 32 t, 76
    exposures, xix–xx, 23–40, 26 t, 44–45, 68–69
    intakes, 28 t
    stay-times, xix, 25–26, 75–76
schoepite, 16
skin contamination, 49–51
source term, 27, 84
special work procedures, 55–57
stay-times
    cancer risks, 33–35
    estimates, 38–39
    Level I exposures, 68–69
    scenarios, 23–26, 75–76
    vehicles, 48–49, 53–55

T
tanks. *see* Abrams tanks
tetravalent compounds, 51
thoracic lymph notes, 20
toxicity
    chemical, xx–xxii, 17–18, 28–31
    combat, 64
    kidneys, 29, 58–59, 67–68
    standards, 63–64
    uranium, 12
training, hazard awareness, xxvi, 70, 79
triangular distributions, 50–51
turret, xvii, 14, 84. *see also* vehicles

U
U.S. Army Center for Health Promotion and Preventive Medicine (USACHPPM), 1–6, 44–45, 55, 56–60, 78
U.S. Army Heavy Metals Office, 4
U.S. Army Medical Command, xvii, 7
U.S. Environmental Protection Agency, 20, 35
U.S. Nuclear Regulatory Commission, xxii, 32–33, 35, 58

V

vehicles. *see also* Abrams tanks; Bradley vehicles
  contamination, 55–56
  exposures, 3–4
  field tests, 14
  perforations, 3, 25, 35 f, 44–45, 66, 77
  unventilated, xix–xx, xxvii, 74
ventilation systems
  Abrams tanks, xvii–xx, 69
  definition, 84
  exposure reduction, xxvii, 14, 30, 56, 77
  Level I exposures, 69–70
  use of, xix–xx, 75
veterans, xxviii, 79

W

weighting factor, 29
wipe tests, 15–17, 49, 52–53
worst-case estimate, 36–40, 39 t, 59
wounds and injuries
  fragment, xxviii, 23, 74, 79
  individual dose effects, 69
  Level I exposures, 3–4